# THE WEATHER REPORT

## LESSON PLANS, WORKSHEETS, AND EXPERIMENTS

### by MIKE GRAF

**FEARON TEACHER AIDS**
A Division of Frank Schaffer Publications, Inc.

This Fearon Teacher Aids product was formerly manufactured and distributed by American Teaching Aids, Inc., a subsidiary of Silver Burdett Ginn, and is now manufactured and distributed by Frank Schaffer Publications, Inc. FEARON, FEARON TEACHER AIDS, and the FEARON balloon logo are marks used under license from Simon & Schuster, Inc.

Illustrator: Duane Bibby

Entire contents copyright © 1989 by Fearon Teacher Aids, 23740 Hawthorne Boulevard, Torrance, CA 90505-5927. However, the individual purchaser may reproduce designated materials in this book for classroom and individual use, but the purchase of this book does not entitle reproduction of any part for an entire school, district, or system. Such use is strictly prohibited.

ISBN 0-8224-7511-1

Printed in the United States of America

1. 15 14 13 12

# Table of Contents

*Introduction*  5

*Acknowledgments*  6

1. What Is Weather?  7
2. Temperature  11
3. Humidity  23
4. Clouds and Fog  33
5. Precipitation  43
6. Wind  57
7. Air Pressure  69
8. Fronts and Storms  85
9. Topography and Weather  101
10. Geography and Climate  111
11. Recording and Forecasting Weather  129
12. Building Your Own Weather Station  137

*Review Test for Weather Unit*  145

*Glossary*  149

*Resources*  150

*Science Equipment Suppliers*  151

*Appendix: Climatic Data for Selected U.S. Cities*  152

*Weather Pen Pals*  160

# Introduction

When you finish *The Weather Report*, not only will your students have learned the why's and how's of weather, but you will also have a school weather station to use throughout the year. Each chapter of *The Weather Report* provides lesson ideas and activity worksheets about one aspect of weather and about the instruments that measure it. You'll find instructions for making simple weather instruments and suggestions for where to obtain more durable ones.

As your students learn how to run a daily weather station, they will also be learning elementary physics, geography, and data analysis. Each chapter summarizes the main concepts behind one aspect of weather and describes demonstrations and experiments to show how it happens. These demonstrations are accompanied by diagrams and illustrations to use with an opaque projector or to copy on the blackboard as you explain the principles of sun, wind, and rain. Reproducible worksheets in each chapter give your students practice in reading weather instruments, charting and mapping weather conditions, and analyzing the influence of terrain and geographical location on weather.

## How to Use This Book

*The Weather Report* can be adapted for students from primary grades through high school. For younger students (grades 3 to 5), focus on the material at the beginning of the book: learning how to read a thermometer, how clouds are formed, how to estimate wind speed and direction. They can report their observations of the weather on a simple felt board chart (pages 129–135). Older students (grade 5 to high school) will be able to do some of the experiments themselves and use the more sophisticated records and charts to analyze and predict weather. Their knowledge of geography will be enhanced by the chapter on geography and climate. Sprinkled throughout each chapter you'll find "Did you know that . . ." boxes; use them to motivate student interest in the corresponding lessons.

The activities, lessons, and experiments in this book do not necessarily have to be completed in the order in which they are presented. Many teachers use the information in this book to supplement the weather chapters in their science text. Others begin a weather unit by teaching how to use the weather instruments and setting up a weather station first and then going back and working through the principles that cause weather.

If you choose to teach a unit on weather, do it when the weather in your area is most varied. The unit can be completed in about a month, depending on how many of the experiments and activities you do. Many of the instruments can remain with the class throughout the year, or the children can take them home to set up their own weather stations.

Once your class knows how to read the weather instruments, weather monitors, or "experts," can do the regular charting, using a minimum of classroom time. Monitors can be switched every couple of weeks with the experienced monitors helping the new ones. The monitors can give two- to three-minute weather reports at the end of each day.

If possible, supplement the instruction with field trips, satellite pictures, class speakers, and weather instruments. Your local National Weather Service station, TV stations, and science catalogs (see page 151) can help you enrich your weather unit.

# *Acknowledgments*

Collecting weather data has been a hobby of mine since childhood. As the years went on, my means of doing this became more sophisticated. I have been fortunate to share this hobby with students, teachers, and camp counselors.

Many people have encouraged me and contributed to my knowledge of teaching weather. I am very grateful to all of them, but special recognition should go to a group of professionals who have helped me in this endeavor: Rich Gerston, naturalist, environmental education camp director, and weather pen pal; Barbara Mahler, professor at California State University, Chico, who provided many resources in teaching weather to children; Tom Loffman, Sacramento TV meteorologist, who provided resources, TV affiliation, and up-to-the-minute weather information; my brother, Danny Graf, who as a college student of meteorology provided ideas, feedback, and enthusiasm for the book; Marion and Max Caldwell, camp directors who allowed me to build and run a weather station at their children's camp in the Santa Cruz mountains; John Poling, professor at California Polytechnic State University (San Luis Obispo), who provided ideas for several weather projects and instruments; and Martha Silva, principal at Branch Elementary School in Arroyo Grande, California, who provided me with the support and the setting for running a weather station with my fourth-grade class. To all of the above and to the many others who also helped, my sincere thanks.

*Mike Graf*

# CHAPTER 1
# WHAT IS WEATHER?

Meteorology, or the study of the atmosphere and its interaction with the earth's surface, began as far back as 340 B.C., when Aristotle began studying such phenomena as clouds, rain, snow, wind, hail, thunder, and hurricanes. In the 1600s more sophisticated observations of weather began with the invention of simple weather instruments, such as the barometer. Today, our observation and prediction of weather and climate is aided by computers and satellites.

Observing and forecasting weather is important to many people. To do their jobs, farmers, pilots, and fishermen must consult weather forecasts every day. Many recreational activities, such as camping, hiking, and picnicking, depend on weather forecasting.

## Concept

Weather includes temperature, clouds and sky conditions, precipitation, wind, and storms.

## Activities

1. **Exciting weather.** Ask your students to describe the most exciting weather event they have ever witnessed.

2. **Fill the bulletin board with weather.** Ask each student to look through magazines and newspapers for weather-related pictures to put up on the bulletin board.

3. **A weather poem.** Hand out the activity worksheet "My Weather Poem," and explain the instructions line by line. Here is an example:

*Cloud*
(the name of anything related to weather)

*You are so fluffy*
(compliment this thing)

*Do you ever want*
(ask it a question)

*to come down?*

*If I were you, I'd say*
(give it some advice)

*hello once in a while.*

Put the illustrated poems on the bulletin board.

4. **The sounds of weather.** A series of environmental records are available in many record stores that include the sounds of weather phenomena such as thunderstorms, blizzards, rain, and wind. Play these sounds for your students to set them thinking about the varieties of weather and how weather affects them.

You might have your students write descriptive paragraphs or short stories about the different kinds of weather they are listening to.

# *Did you know that . . .*

in Europe, stores sell tree frogs in small glasses to be used as weather forecasters? Although there is no scientific proof, it has long been a popular adage that when frogs croak more than usual or when ducks quack more than usual, rain is on its way. The assumption that these animals can predict rain seems to be based on their close association with water.

## Concept

Many plants and animals are sensitive to changes in air pressure and humidity and react to a change in the weather.

## Activity

**Natural weather indicators.** Give each student the activity worksheet "Nature's Weather Forecasters" and have them work through it before discussing the answers in class. You might also ask students to share other weather signs they have heard of—or to make up their own.

ANSWER KEY:
**Nature's Weather Forecasters**

(All answers may vary.) **1.** Possibly true; rain may wash away the trail scent, so ants will stay closer together to keep the scent. **2.** Possibly true; ants try to protect their nests from the moisture. **3.** False; storage depends on how many nuts are available. **4.** False. **5.** False. **6.** Possibly true; sunflowers may raise their heads to get the last bit of sunlight, or in reaction to increased moisture. **7.** False. **8.** False. **9.** Possibly true; a sense of air pressure change may induce restlessness. **10.** Possibly true; animals will spread out when there is no storm danger or need for warmth and protection. **11.** False; the croaking of frogs is a mating call. **12.** Possibly true; deer and elk move to lower ground to get away from a dangerous area to a more protected one. **13.** False. **14.** False.

NAME _____

# MY WEATHER POEM

_____
(the name of anything related to weather)

_____
(compliment that thing)

_____
(ask it a question)

_____
(give it some advice)

Draw a picture here to go with your poem.

*The Weather Report © 1989*

WHAT IS WEATHER?    9

NAME _____

# NATURE'S WEATHER FORECASTERS

Many plants and animals react to a change in the weather. They are sensitive to changes in air pressure and humidity. Which of the following are true statements about how plants or animals react to weather? Mark each statement true or false.

T   F

____ ____ 1. In fair weather, ants move scattered about. In rainy weather, they move in a single-file line.

____ ____ 2. When ants sense rain, they close up their nests.

____ ____ 3. If a squirrel stores many nuts, expect a severe winter.

____ ____ 4. If you slap away a fly on your nose and it comes back, a storm is coming.

____ ____ 5. If a groundhog sees its shadow on February 2, expect six more weeks of winter.

____ ____ 6. When sunflowers raise their heads, it is a sign of rain.

____ ____ 7. When a cat scratches itself on a fence, it is a sign of rain.

____ ____ 8. Flies bite excessively before stormy weather.

____ ____ 9. Horses race around before a violent storm.

____ ____ 10. When sheep go into the hills and scatter, expect nice weather.

____ ____ 11. Frogs croak more than usual before a storm.

____ ____ 12. Deer and elk come down from the mountains at least two days before a storm.

____ ____ 13. Bluebirds chatter when it's going to rain.

____ ____ 14. Birds on a telephone wire indicate the coming of rain.

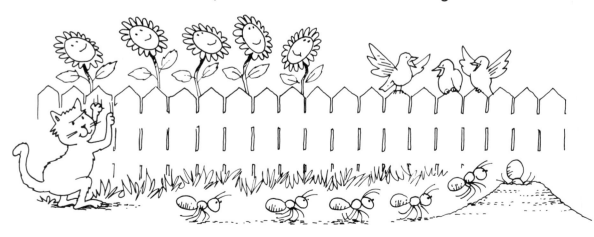

10    WHAT IS WEATHER?

# CHAPTER 2
# TEMPERATURE

Temperature is a measure of the average speed of the random motions of air molecules. The faster the molecules move, the higher the temperature. Temperature is determined by many factors, especially sunlight and wind. (See pages 85–127 for the effect of latitude, altitude, and oceans on temperature; if your students are in the upper grades, you may wish to include those lessons here.)

## Concept

Temperature can be measured.

## Activities

1. **Finger thermometers.** Set up groups of three bowls of hot (not scalding), lukewarm, and cold water. Have students pretend their fingers are thermometers and put a finger in each bowl. Ask them to describe the different temperatures they feel. Use their answers to introduce the concept of numerical degrees to describe how hot or how cold something is.

2. **Reading a thermometer: demonstration.** Use page 15, "Reading a Thermometer," on an opaque projector. Cut out the "mercury" strip and have the students read various temperatures.

    You might also explain how a thermometer works: It is a tube filled with liquid that rises as it gets hotter.

3. **Reading a thermometer: practice.** Give the students the worksheet "How Hot? How Cold?" Have them answer the questions.

4. **Reading a maximum-minimum thermometer: demonstration.** Use page 17, "Reading a Maximum-Minimum Thermometer," on an opaque projector. Cut out the "mercury" and "magnet" strips and have the students read various temperature ranges. The mercury pushes the magnets to the extreme low or high temperature for the day and then recedes, leaving the magnets in place, until they are reset. Note that the scale on the low side is read from the top down, and the scale on the high side is read from the bottom up, like a regular thermometer. The high and low temperatures are indicated by the position of the magnet in each column.

5. **Reading a maximum-minimum thermometer: practice.** Give the students the activity worksheet, "Highs and Lows." Have them fill in the high and low temperatures.

    Once they have mastered reading a regular thermometer and a maximum-minimum thermometer, have your students record a week's high and low temperatures on the chart on the worksheet, using your school's maximum-minimum thermometer. (Don't forget to reset the thermometer at the end of

# *Did you know that...*

a cricket is a great thermometer! In the evening when crickets are in the cool grass, count the number of chirps they make in 14 seconds. Then add 40. That will be the temperature in their location (the temperature may be different where you are standing).

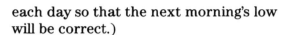

each day so that the next morning's low will be correct.)

6. **Making a thermometer.** See page 19. If your students are in the upper grades, duplicate these instructions and let them make their own thermometers.

ANSWER KEY:
**How Hot? How Cold?**

A. **1.** 85°.  **2.** 71°.  **3.** 60°.  **4.** 52°.
**5.** 38°.

B.

ANSWER KEY:
**Highs and Lows**

|     | Low | High |
|-----|-----|------|
| 1.  | 42° | 44°  |
| 2.  | 49° | 58°  |
| 3.  | 52° | 65°  |
| 4.  | 58° | 73°  |
| 5.  | 62° | 79°  |

THE WEATHER REPORT

## Concept

Temperature variations are caused by many factors, including sun, wind, shade, surfaces, colors, and reflections.

## Activities

1. **Microclimates.** Provide students with a sample map of your school on 12″ × 18″ paper, or have them draw one. Draw a "master" map on the board (see the sample map below). Assign partners to take a thermometer to various locations and record the temperature. To get an accurate reading, they should leave the thermometer in place for five minutes (remind them not to hold or breathe onto the thermometer). Plot all the temperatures on the maps. Discuss reasons for the variations in temperature. For example, thermometers in the sun read the movement of the air molecules (temperature) as well as of the radiant energy of the sun. To find the actual temperature, a thermometer should be placed in the shade, where it will read only the movement of the air molecules.

2. **How color influences temperature.** Put a thermometer in a coffee can that is painted white, and put another in a coffee can that is painted black. Put both out in the sun. Check the temperatures every five minutes and record them. Discuss the results. The thermometer in the black can will be warmer because black absorbs light and white reflects it. NOTE: A black surface absorbs more heat during the day and gives off more heat at night than a white surface does. You can use a maximum-minimum thermometer to demonstrate this theory.

3. **Seasonal temperature variations.** Discuss the seasonal temperature ranges in your area. (See Appendix for state-by-state temperature ranges.) How cold does it get on a winter night? How hot does it get on a summer afternoon? Locate those temperatures on a thermometer. How do the sunlight and the wind differ in different seasons? How do those differences affect temperature?

temperatures at 2:00 p.m.

**4. Reading a weather map: temperature.** Introduce the worksheet "Mapping Temperatures" by telling the students that, in the course of this weather unit, they will learn how to read and make weather maps like those they see in the newspaper and on television. Discuss the example for San Francisco temperatures on the worksheet, making sure students understand that the high is written above the line and the low below it. For the high and low temperatures for your city, you might want to use the temperatures recorded on the worksheet "Highs and Lows" (page 18). Older students might also mark temperatures for a few other cities, using newspaper weather information.

**5. Global temperature variations.** Duplicate the worldwide weather report from your newspaper, or show it on an opaque projector. Compare the temperatures around the world with the current high and low in your area. Where is it hotter? How much hotter is the hottest place? Where is it colder? How much colder is the coldest place? Where are the temperatures similar to your area?

To explore global temperature variations further, turn to pages 85–127 for lessons on the effects of altitude, latitude, and oceans on temperature.

## ANSWER KEY:
### Mapping Temperatures

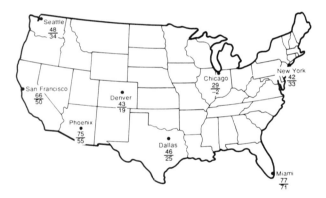

## ANSWER KEY:
### Temperature Review Test

1. True.  2. True.  3. False.  4. True.
5. D.  6. A. 24°.  B. 35°. C. 47°.
7. A. 60°.  B. 31°.  C. 48°.

14    THE WEATHER REPORT

# READING A THERMOMETER

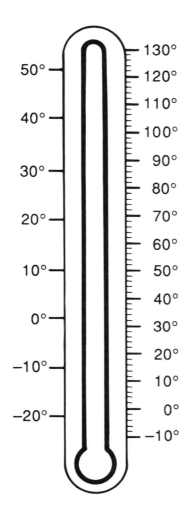

---

Cut out this strip and use it as "mercury" to demonstrate various temperatures.

*The Weather Report* © 1989

TEMPERATURE

NAME _____

# HOW HOT? HOW COLD?

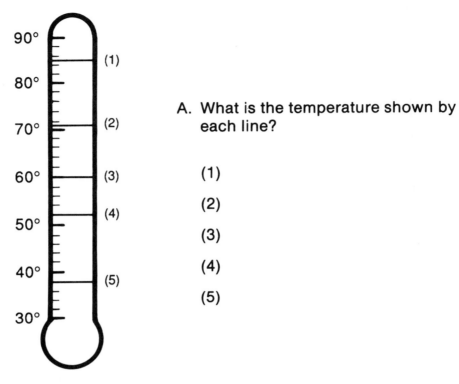

A. What is the temperature shown by each line?

(1)

(2)

(3)

(4)

(5)

B. Draw a line across the thermometer to show each of the following temperatures. Number each temperature line.

(1) 39°

(2) 44°

(3) 56°

(4) 68°

(5) 80°

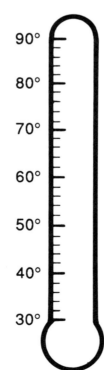

*The Weather Report* © 1989

16   TEMPERATURE

# READING A MAXIMUM-MINIMUM THERMOMETER

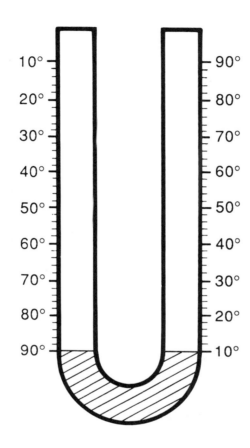

---

Cut out these strips and use them as "mercury" and "magnets" to demonstrate various temperatures.

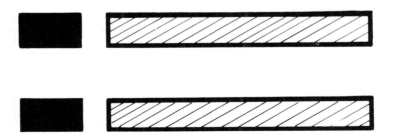

NOTE: The magnet touches the mercury when the thermometer is reset; once the mercury recedes from the high and low readings, it separates from the magnet.

TEMPERATURE

NAME _____

# HIGHS AND LOWS

Each line on this maximum-minimum thermometer represents a high or low temperature recorded by the magnet. What are the highs and lows shown by each pair of lines?

|    | LOW | HIGH |
|----|-----|------|
| 1. | 42° | 44°  |
| 2. |     |      |
| 3. |     |      |
| 4. |     |      |
| 5. |     |      |

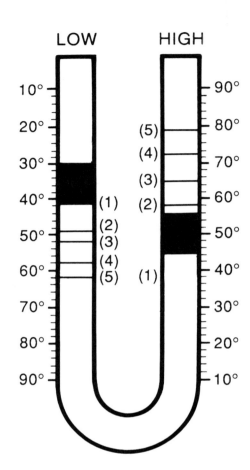

## The Week's Highs and Lows

Use a maximum-minimum thermometer to record the high and low temperatures for each day this week—as well as the current temperature and the time the temperature is read each day.

|           | HIGH TEMPERATURE | LOW TEMPERATURE | CURRENT TEMPERATURE | TIME OF READING |
|-----------|------------------|------------------|---------------------|-----------------|
| Monday    |                  |                  |                     |                 |
| Tuesday   |                  |                  |                     |                 |
| Wednesday |                  |                  |                     |                 |
| Thursday  |                  |                  |                     |                 |
| Friday    |                  |                  |                     |                 |

18   TEMPERATURE

NAME _____

# MAKING A THERMOMETER

## Materials

Small plastic container with lid
  (film containers work well)
Clear plastic straw
Food coloring or ink
  (to color water)
Index card

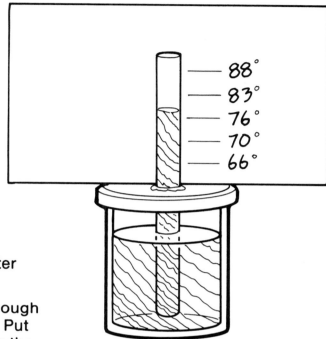

## Directions

1. Fill container two-thirds full of water and add coloring.

2. Make a hole in the lid just large enough for the straw. Put lid on container. Put straw into lid. Don't let straw touch the bottom of container.

3. Seal around the straw with glue. Seal around the lid with glue. The thermometer needs to be airtight.

4. Once the colored liquid is in the container and it is all sealed and dry, blow into the straw until the water level rises to just above the cap or lid. Then it will be ready.

5. Tape index card behind the straw.

6. Take your thermometer to various locations with a real thermometer. Mark the degrees on the index card as the colored water moves up the straw. Stay in each location long enough for the liquid to settle.

NOTE: This thermometer is temporary. Once the air pressure changes, it loses its accuracy. It is good for at least a day, though.

*The Weather Report* © 1989

TEMPERATURE    19

NAME _____

# MAPPING TEMPERATURES

Record the temperature information on the weather map. San Francisco's high and low temperatures are already noted. Use it as an example.

1. San Francisco's weather is mild, with a high of 66° and a low of 50°.
2. Seattle is seasonably cool, with the temperature ranging between 48° and 34°.
3. Phoenix, as always, is warmer than most of the country; its high is 75° and its low is 55°.
4. It will freeze in Denver tonight, with a low of 19°, down from a high of 43°.
5. Dallas, with a high of 46° and a low of 25°, is nearly as cold.
6. People are shivering in Chicago, where the high is only 29° and the low is a numbing –2°.
7. New York's temperatures will stay just above freezing, with a high of 42° and low of 33°.
8. Miami is the place to really get out of the winter cold, with a balmy high of 77° and a not much lower low of 71°
9. Now locate your city on the map, label it, and show its high and low temperatures for today.

TEMPERATURE

*The Weather Report* © 1989

NAME _____

# Temperature Review Test

1. True ____ or False ____ . Animals may react unusually if a storm is coming.

2. True ____ or False ____ . When the temperature increases, the air molecules speed up.

3. True ____ or False ____ . The basketball courts are probably the coolest area of a school.

4. True ____ or False ____ . When you record temperatures on a map, the high temperature goes above the line and the low temperature goes below the line.

5. Which of the following causes temperature variations?

   ____ A. Wind      ____ C. Reflections

   ____ B. Sun       ____ D. All of the above

6. What are the temperatures shown on this thermometer?

   A. _____

   B. _____

   C. _____

7. On this maximum-minimum thermometer,

   A. what is the high temperature?

   _____

   B. what is the low temperature?

   _____

   C. what is the current temperature?

   _____

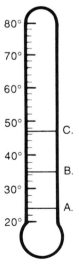

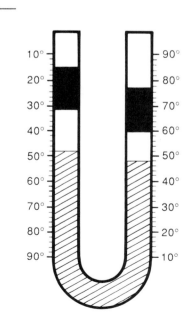

TEMPERATURE 21

# CHAPTER 3
# HUMIDITY

Humidity is the amount of water vapor in the air. Water becomes vapor through evaporation. Evaporation is caused by heat and wind.

Water vapor becomes liquid again through condensation. Condensation is caused by cooling. Condensation occurs when air reaches its saturation point. The saturation point depends on the temperature; warm air can hold more water vapor than cold air can. The temperature at which water vapor will condense is called the dew point.

Water vapor that condenses on objects is visible as dew or frost. Water vapor that condenses in the air near the ground is fog. At higher altitudes water vapor condenses to become clouds. (See next chapter for fog and clouds.)

## Concept

The effects of humidity can be observed.

## Activities

1. **The salt won't shake and the sugar clumps.** Try to pour salt or sugar in a humid environment. It's hard to pour and forms small clumps, because water vapor has condensed on the salt or sugar crystals and has made them stick together.

2. **Pine cones close up.** Put a pine cone outside where you can observe it from time to time. How does it change when the humidity increases? It closes up in moist weather to protect the seeds.

3. **Leaves crunch.** Dry leaves are crunchy when the humidity is low. They are flexible when the humidity is high. Dampen old, dry leaves to demonstrate.

## Concept

Evaporation is caused by heat (usually from the sun) and wind.

## Activities

1. **The sun causes evaporation.** Place 10 drops of water in each of two pie tins. Put one pie tin in the shade and one in the sun. Observe both immediately.

2. **The wind causes evaporation.** Wet two spots on the chalkboard with a wet sponge. Fan one spot vigorously. Which spot dries more quickly?

## Concept

Water vapor rises because it has been heated.

## Activity

**Water vapor is visible.** Set up a slide projector on a table and direct the beam of light right over a bowl of very hot water. Turn off the room lights and observe. What are seen rising are tiny water droplets that have condensed out of rising warm air that is rich in water vapor.

# Did you know that...

it is more efficient to water your lawn in the evening than in the afternoon? On a warm day, up to 50 percent of the water is lost to evaporation when the sun is shining overhead.

## Concept

Warm air holds more water vapor than cool air.

## Activities

1. **Evaporation in warm air and in cold air.** For this demonstration you will need two identical glass jars with lids, two 1-inch squares of cloth, and a needle and thread.

   Put one jar, uncovered, in sunlight or near a heating vent for about 30 minutes. Put the other jar in the refrigerator for the same amount of time. Meanwhile, use the needle to pull through a 4-inch length of thread at the midpoint of one side of each square. Tie the thread to the cloth, leaving one tail as long as possible. Tape the end of this long tail to the underside of a jar lid. Wet both squares of cloth and gently squeeze out excess water.

   Now remove the jars from their warm and cold spots. Put a lid on each jar so the cloth square hangs into the jar. Here is what your jar should look like.

   Now put the jars back into the warm and cold spots they came from and leave them there for another 30 minutes. Then remove the jars and feel the cloths. Which cloth is drier? Where did the water from the cloths go? Did the warm air or the cold air hold more water vapor? The warm jar should have a dry cloth due to the higher evaporation rate.

2. **The warmer the air, the more water vapor it can hold.** Use page 28, "Air Temperature and Water Vapor Capacity," in an opaque projector to explain this concept to your students. The heavy line shows the saturation absolute humidity, or the point at which the air can hold no more water vapor. The dotted lines compare the amount of water vapor that air at 50° can hold, with the amount of water vapor that air at 95° can hold—almost four times as much.

## Concept

Humidity can be measured.

## Activities

1. **Hair hygrometer.** Cut a 12" × 12" square of *heavy* cardboard. Cut a 6-inch-long arrow from thin cardboard. Punch two holes near the square end of the arrow. Into the lower hole, tie one end of a hair that is at least 10 inches long (horse hair, if it's available, is sturdier than

24    THE WEATHER REPORT

human hair). *Loosely* attach the arrow to the cardboard square by putting a tack through the top hole. Push another tack part way into the cardboard square 6 inches below the bottom hole in the arrow. Pull the hair until it's straight and then tie it onto the lower tack and push the tack all the way into the cardboard. The hair will expand or contract slightly depending on the amount of moisture in the air. Calibrate the movement of the pointed end of the arrow with a real hygrometer. Mark these readings on the cardboard square. NOTE: It's hard to make this experiment successful, but it's an interesting idea to share with your students.

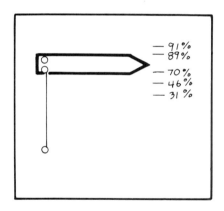

2. **Milk-carton hygrometer.** For instructions, see page 140 of the chapter "Building Your Own Weather Station." The less humidity in the air, the faster the water in the wick of the wet-bulb thermometer evaporates and the more heat is carried away by evaporation.

   To interpret the readings from your milk-carton hygrometer, refer to the table on the worksheet "Relative Humidity." You might post a larger version of this table in the classroom for easy reference.

3. **Relative humidity.** Relative humidity is the comparison of the amount of moisture present in the air and the amount of moisture that it could hold. A relative humidity reading of 100 percent means that the air is completely filled with water vapor. A relative humidity reading of 0 percent means that there is no water vapor in the air. Relative humidity measures how humid or how dry the air feels; absolute and specific humidity are measures of the quantity of water in a given amount of air.

   After you make the milk-carton hygrometer, hand out the activity worksheet "Relative Humidity" to give your students practice in using a wet-bulb-dry-bulb hygrometer to find relative humidity.

4. **Heat index.** To discuss how various humidity levels affect people, show page 30, "Heat Index," on an opaque projector. If you live in an area where summers are humid, you'll know from first-hand experience that the higher the humidity, the more uncomfortable a person will feel.

ANSWER KEY:
**Relative Humidity**

1. 55%.  2. 6%.  3. 55%.  4. 92%.
5. #2, 6%.  6. #4, 92%.

## Concept

Water vapor condenses when the air is cooled.

## Activities

1. **Condensation (I).** For this demonstration, you will need two identical heatproof jars. Fill one jar a quarter full of water. Place the other jar on top of it so the mouths of the two jars are touching. Tape the jars together at their mouths. Place the pair of jars in a pan of water, with the water-filled jar beneath the empty jar. Put a tray of ice on the empty jar. Heat the water in the pan to boiling and keep it boiling. What happens in the top jar? This may be easier to see if you turn the room lights off and shine a flashlight into the top jar. The warm air rises and condenses as it meets the cooler air.

HUMIDITY 25

2. **Condensation (II).** Put one drop of water in a jar with a medicine dropper. Cover the jar and seal it tightly. Put the jar under a lighted lamp and observe it after 15 minutes. What happened to the drop of water?

   Now put the same jar in a bowl of ice cubes for 15 minutes. Then dry the outside of the jar. Can you see the water inside now? Why? Condensation occurs when the air is cooled to a point where the air can hold no more water vapor, so it condenses into water and becomes visible.

## Concept

Warm air holds more water vapor than cool air. When warm and cool air meet, the warm air is cooled and becomes saturated, or 100 percent full of water. When water vapor condenses on objects, it becomes dew or frost.

## Activities

1. **Making dew.** This demonstration requires two identical clean, dry jars. Put ice and water into one jar, and then cover it. Cover the empty jar as well. Place the two jars near one another and observe. After a while you'll notice that the jar with the ice has water condensing on it. The cooler the air is, the less water it can hold, and therefore it becomes more quickly saturated. The jar without ice remains dry because the air was not cooled. At night dew forms first on the ground, because that is where the air is coolest.

2. **Dew and frost.** This demonstration uses two tin cans. In one can put only ice. In the other, put ice and rock salt. The salt cools the temperature further. Frost will form on the outside of the salted can; dew will form on the outside of the other.

## Concept

The temperature at which water vapor will condense is called the dew point. (If this temperature is below freezing, it is called the frost point.) The dew point and the frost point are measures of humidity.

## Activities

1. **Dew-point temperature (I).** Put crushed ice in a tennis ball can that has been weighted down with metal weights or rocks in the bottom. Put the tennis ball can on a 1-inch-high block of wood inside a coffee can that has been painted flat black. Put the thermometer in the coffee can. When dew forms, record the temperature. This is the dew point.

2. **Dew-point temperature (II).** Fill a shiny metal can half full of water. Tape a thermometer inside the top of the can and add a small amount of ice. (Don't let the thermometer point touch the water.) Stir the ice in the water and watch the sides of the can. As soon as dew forms on the outside of the can, read the temperature. This is the dew point, or the temperature at which the air becomes saturated. Note the room air temperature and the dew-point temperature. The bigger the difference between the two temperatures, the lower the humidity.

26  THE WEATHER REPORT

# *Did you know that . . .*

you can give yourself a halo? On a cool morning when there is dew on the grass, stand with your back to the sun. You should see a bright area (from the sun's reflection on the dewdrops), which is called a *heiligenschein* (the German word for halo), around the shadow of your head.

ANSWER KEY:
**Humidity Review Test**

**1.** Saturation.   **2.** Dew.   **3.** Hygrometer.
**4.** Evaporation.   **5.** Humidity.
**6.** Condensation.   **7.** False.   **8.** True.

# AIR TEMPERATURE AND WATER VAPOR CAPACITY

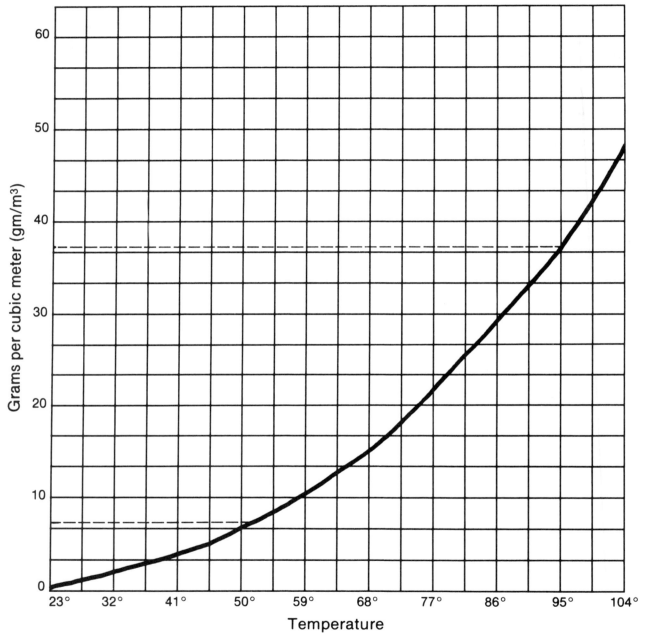

28  HUMIDITY

NAME _____

# RELATIVE HUMIDITY

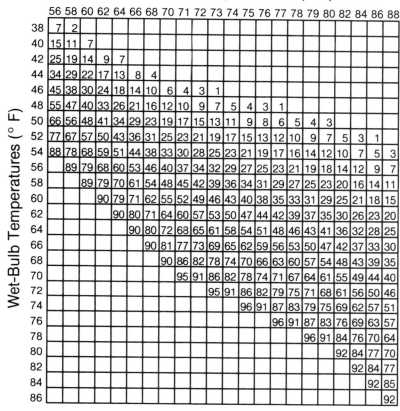

Use the table above to fill in the relative humidity for these readings.

|    | Wet-Bulb Reading | Dry-Bulb Reading | Relative Humidity |
|----|------------------|------------------|-------------------|
| 1. | 60° | 70° | ____ % |
| 2. | 46° | 70° | ____ % |
| 3. | 70° | 82° | ____ % |
| 4. | 80° | 82° | ____ % |

5. At which humidity reading of the four above would it be driest? _____

6. At which reading of the four above would you expect rain soon? _____

HUMIDITY 29

# HEAT INDEX

### Relative Humidity (%)

| Air Temp (°F) | 0 | 5 | 10 | 15 | 20 | 25 | 30 | 35 | 40 | 45 | 50 | 55 | 60 | 65 | 70 | 75 | 80 | 85 | 90 | 95 | 100 |
|---|---|---|---|---|---|---|---|---|---|---|---|---|---|---|---|---|---|---|---|---|---|
| 140 | 125 | | | | | | | | | | | | | | | | | | | | |
| 135 | 120 | 126 | | | | | | | | | | | | | | | | | | | |
| 130 | 117 | 122 | 131 | | | | | | | | | | | | | | | | | | |
| 125 | 111 | 116 | 123 | 131 | 141 | | | | | | | | | | | | | | | | |
| 120 | 107 | 111 | 116 | 123 | 130 | 139 | 142 | | | | | | | | | | | | | | |
| 115 | 103 | 107 | 111 | 115 | 120 | 127 | 135 | 143 | 151 | | | | | | | | | | | | |
| 110 | 99 | 102 | 105 | 108 | 112 | 117 | 123 | 130 | 137 | 143 | 150 | | | | | | | | | | |
| 105 | 95 | 97 | 100 | 102 | 105 | 109 | 113 | 118 | 123 | 129 | 135 | 142 | 149 | | | | | | | | |
| 100 | 91 | 93 | 95 | 97 | 99 | 101 | 104 | 107 | 110 | 115 | 120 | 126 | 132 | 133 | 144 | | | | | | |
| 95 | 87 | 88 | 90 | 91 | 93 | 94 | 96 | 98 | 101 | 104 | 107 | 110 | 114 | 119 | 124 | 130 | 136 | | | | |
| 90 | 83 | 84 | 85 | 86 | 87 | 88 | 90 | 91 | 93 | 95 | 96 | 98 | 100 | 102 | 106 | 109 | 113 | 117 | 122 | | |
| 85 | 76 | 79 | 80 | 81 | 82 | 83 | 84 | 85 | 86 | 87 | 88 | 89 | 90 | 91 | 93 | 95 | 97 | 99 | 102 | 105 | 108 |
| 80 | 73 | 74 | 75 | 76 | 77 | 77 | 78 | 79 | 79 | 80 | 81 | 81 | 82 | 83 | 85 | 86 | 86 | 87 | 88 | 89 | 91 |
| 75 | 69 | 69 | 70 | 71 | 72 | 72 | 73 | 73 | 74 | 74 | 75 | 75 | 76 | 76 | 77 | 77 | 78 | 78 | 79 | 79 | 80 |
| 70 | 64 | 64 | 65 | 65 | 66 | 66 | 67 | 67 | 68 | 68 | 69 | 69 | 70 | 70 | 70 | 70 | 71 | 71 | 71 | 71 | 72 |

Air Temperature (°F)

| Heat Index | Possible Heat Disorders for People in Higher-Risk Groups |
|---|---|
| 130° or higher | Heatstroke or sunstroke highly likely with continued exposure |
| 105°–130° | Sunstroke, heat cramps, or heat exhaustion likely, and heat-stroke possible with prolonged exposure or physical activity |
| 90°–105° | Sunstroke, heat cramps, and heat exhaustion possible with prolonged exposure or physical activity |
| 80°–90° | Fatigue possible with prolonged exposure or physical activity |

*The Weather Report* © 1989

NAME _____

# Humidity Review Test

Fill in the blank in sentences 1-6 with one of the words below.

    humidity        evaporation        condensation

    saturation        dew        hygrometer

1. _____ is when the air can hold no more water vapor.

2. _____ is water vapor that has condensed onto objects near the ground.

3. _____ is an instrument for measuring humidity.

4. _____ is the process of water turning into vapor.

5. _____ is the amount of water vapor in the air.

6. _____ is the process of water vapor turning into liquid.

7. True ___ or False ___ . The best time to water the law is mid-afternoon.

8. True ___ or False ___ . Warm air holds more water than cool air.

HUMIDITY    31

# CHAPTER 4
# CLOUDS AND FOG

Water vapor that condenses in the air near the ground is fog. When water vapor condenses above the ground, clouds form. Specks of dust can provide nuclei around which water will condense.

Because of variations in air temperature and wind speed at different altitudes, the clouds formed at different altitudes have different patterns. These various shapes of clouds are used to predict the weather.

## Concept

Clouds and fog are made when water vapor condenses around specks of dust or smoke.

## Activities

1. **Dust in the air.** This demonstration will show your students that there are plenty of particles in the air around which clouds can form.

    Cut out a circle of white paper to fit into the bottom of a pie tin. Coat the paper with petroleum jelly and weight the paper down with one or two small rocks (or use double-stick adhesive tape). Place the tin on the windowsill for a day or two. You'll be surprised at how much dust it collects.

2. **Making fog.** Put ice in a pie tin on top of a jar containing very warm or boiling water. What happens? To see the result most clearly, you might turn off the room light and shine a flashlight into the jar. When warm and cool air meet, the air becomes saturated and fog is formed.

    Fog forms when the air is much cooler near the ground, because the water vapor does not have to rise very far before it condenses. Fog over land is sometimes called valley fog. The fog that occurs over water that is warmer than the air (such as that in a heated swimming pool in the winter) is often called evaporation fog.

3. **Breathing clouds.** On a cool morning, notice what happens when you breathe out. The warm, moist air from your mouth meets the cool outside air and becomes saturated, forming a small puff of fog.

4. **Simple cloud-making (I).** Warm a narrow-necked bottle by shaking hot water in it. Put an ice cube on top of the bottle. Because this demonstration does not add particles of dust or smoke that act as nuclei to help the water vapor condense more visibly, you will have to turn off the room lights and shine a flashlight into the bottle to see the cloud.

    Another variation of this experiment is simply to fill a narrow-necked bottle with 1 to 2 inches of hot water and put the ice cube on top of the bottle.

5. **Simple cloud-making (II).** Wet the inside of a glass milk bottle with warm

33

water. Drop a lighted match into the bottle. (The match will go out, but it will leave some smoke, which provides many tiny carbon particles around which the water droplets can condense.) Stand with your back toward the light (to see better) and *quickly* blow air into the bottle. A cloud should form.

6. **Fancy cloud-making.** This demonstration requires a water-cooler bottle with a rubber stopper, a length of glass tubing, a length of rubber tubing, and an air pump.

   Put the glass tubing into the rubber stopper and fit the rubber tubing onto the end of the glass tubing. Attach the air pump to the rubber tubing. Pour a little water into the bottle. Put a smoking match inside the bottle (the smoke particles provide the nuclei around which the water vapor will condense). Quickly put in the stopper, and pump the air pump a few times. Wait a few seconds and then loosen the stopper to let cool air in. A cloud will form.

   Pumping in air will pressurize and temporarily heat the air in the jar. Letting cool air in will quickly depressurize and cool the air in the jar. This is a process called *adiabatic cooling*. The cooler air allows the cloud to form.

7. **Cloud-making diagram.** To illustrate the process of warm, moist air (holding water vapor) rising and meeting a layer of cooler air to form clouds, use the page "Where Clouds Come From" in an opaque projector or duplicate it to hand out to your students.

   Often, cumulus clouds disappear and re-form in the same spot on the windward side of a mountain. Clouds will dissipate and move along with the wind and then surface heating will again occur, causing the warm air to rise up the mountain and create more clouds in the same spot.

## Concept

Clouds take different shapes at different altitudes. Clouds are identified according to shape.

## Activities

1. **Names of clouds.** Ask students to think of all the words they can to describe clouds. As they call them out, write them on the board, organizing them (insofar as possible) into four columns according to the four main cloud words: (1) feathery, wispy (cirrus); (2) fluffy, piled up (cumulus); (3) sheetlike, layered (stratus); (4) rainy, gray (nimbus). Then hand out the worksheet "Cloud Words." Go over the basic words with your students. Have them write the descriptions of the major cloud types by combining the basic words. You might start them out with an example: cirrocumulus clouds are feathery, piled-up clouds, for instance.

2. **Cloud categories.** Duplicate the page "Cloud Categories" and pass it out to your students. Use it to explain the kinds of clouds that appear at various elevations and to explain how clouds are associated with various kinds of weather (this will be explored further in the "Fronts and Storms" chapter).

34   THE WEATHER REPORT

Cloud elevations can be estimated in relation to hills and other landmarks and by the speed of the clouds' movement: the faster they cross the sky, the lower the clouds.

3. **Identifying clouds.** Hand out the worksheet "Name That Cloud." Students should refer to the page "Cloud Categories" to help them label the clouds.

   Give them additional practice in identifying clouds by asking them to keep a record of the clouds that they see during a week (including the time of observation and related weather). If it's a time of year when there is not much cloud variety in your area, you could ask them to look through books and magazines for pictures of different kinds of clouds. Have the students show their examples on the opaque projector while the rest of the class tries to identify the cloud types.

4. **Cloud chart.** An excellent, inexpensive poster-size cloud chart (or set of cloud charts) can be purchased from Cloud Chart, Inc., P.O. Box 21298, Charleston, SC 29413.

5. **Make your own cloud chart.** This art project will require 12″ × 18″ sheets of construction paper and cotton balls. It illustrates four basic cloud types.

   Fold the construction paper into quarters, and draw lines where the fold marks are. Make ½-inch marks up the left-hand margin of each of the four quarters. Each mark will represent 5,000 feet. Label the marks from 5,000 feet to 55,000 feet. In each quarter, put the name of the cloud: cirrus, cumulus, stratus, nimbostratus. (Or make a cotton picture of a cumulonimbus cloud, showing its nimbus, cumulus and cirrus layers.) Then write in the typical weather associated with each kind of cloud. Use the pictures in this book and other reference books to shape the cotton correctly for each kind of cloud. You can color the cotton with a black felt-tipped pen to give the nimbostratus a gray color. Glue the clouds in place at their proper altitude, referring to the page "Cloud Categories."

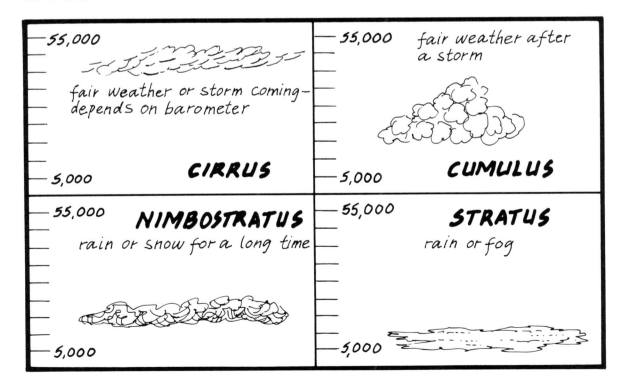

CLOUDS AND FOG

ANSWER KEY:
**Cloud Words**

**1.** feathery clouds. **2.** feathery, piled-up clouds **3.** feathery clouds in sheets (or stretched out). **4.** high, piled-up clouds. **5.** high clouds in sheets (or stretched out). **6.** rain clouds in sheets (or stretched out). **7.** clouds in sheets (or layers or stretched out). **8.** piled-up clouds in sheets (or stretched out). **9.** piled-up clouds. **10.** piled-up rain clouds.

ANSWER KEY:
**Name That Cloud**

**1.** Cirrus. **2.** Cirrocumulus. **3.** Cirrostratus. **4.** Altocumulus. **5.** Altostratus. **6.** Nimbostratus. **7.** Stratus. **8.** Stratocumulus. **9.** Cumulus. **10.** Cumulonimbus.

ANSWER KEY:
**Clouds and Fog Review Test**

**1.** Nimbus: rain. Cirrus: fair, with rain possible within two days. Cumulus: fair weather. Cumulonimbus: thundershowers.
**2.**

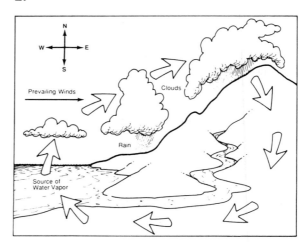

**3.** Stratus: low. Cirrus: high. Cirrocumulus: high. Cumulus: low to middle. Altostratus: middle.

# WHERE CLOUDS COME FROM

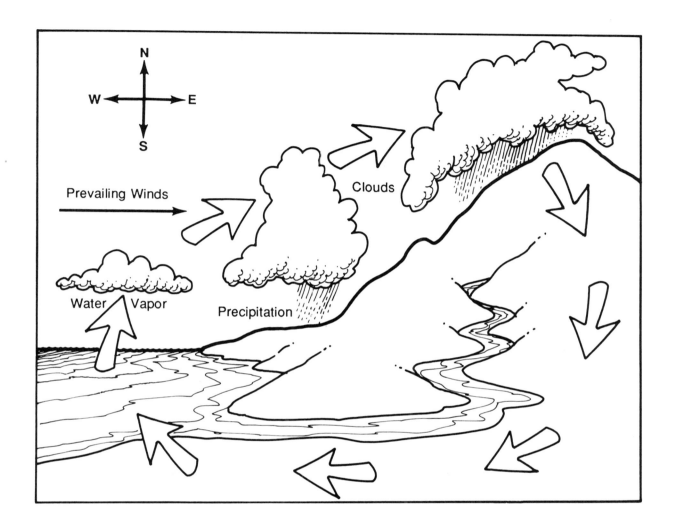

1. The water in lakes and oceans evaporates.
2. Because it is warm, the water vapor rises.
3. When the warm, moist air meets air that is cold enough to cool it to its dew point, adiabatic cooling takes place and clouds form.
4. When the water vapor turns into droplets that are too heavy to be held aloft by wind currents, precipitation occurs.
5. The precipitation runs off into rivers, which carry the water back to lakes and oceans.

CLOUDS AND FOG

NAME _____

# CLOUD WORDS

The names of clouds come from Latin words that describe their appearance. Here are the five words that are used alone or in combination to name the basic cloud types.

**cirrus**  feathery
   (from Latin *cirrus*, meaning "curl, filament, tuft")

**cumulus**  piled up
   (from Latin *cumulus*, meaning "heap, mass")

**stratus**  sheet
   (from Latin *stratus*, meaning "stretched out, extended")

**nimbus**  rain
   (from Latin *nimbus*, meaning "heavy rain; rain cloud")

**alto**  high
   (from Latin *altus*, meaning "high")

The names of the ten basic cloud types use these words alone or in combination. See if you can figure out what each of these clouds looks like from its name.

1. Cirrus clouds are _____

2. Cirrocumulus clouds are _____

3. Cirrostratus clouds are _____

4. Altocumulus clouds are _____

5. Altostratus clouds are _____

6. Nimbostratus clouds are _____

7. Stratus clouds are _____

8. Stratocumulus clouds are _____

9. Cumulus clouds are _____

10. Cumulonimbus clouds are _____

*The Weather Report © 1989*

38   CLOUDS AND FOG

NAME _____

# CLOUD CATEGORIES

| Cloud Type (and symbol) | Description | Composition | Height | Related Weather | Forecast |
|---|---|---|---|---|---|
| **High Clouds:** | | | | | |
| Cirrus | delicate, threadlike | ice crystals | above 20,000 ft. (often higher than 35,000 ft.) | | 1st sign of approaching storm or weather change |
| Cirrocumulus (rare) | small tufts of cotton | ice crystals | above 20,000 ft. | quiet winter weather | |
| Cirrostratus | thin sheet; causes halo around sun or moon | ice crystals | above 20,000 ft. | | rain or snow likely within 24 hours |
| **Middle Clouds:** | | | | | |
| Altocumulus | unconnected piles or layers piled together | water droplets | 6,000 to 20,000 ft. | quiet winter weather | summer thunderstorm |
| Altostratus | smooth white or gray sheet; sun may be seen through clouds | water droplets | 6,000 to 20,000 ft. | maybe light rain or snow | rain or snow likely in 6 to 8 hours |
| Nimbostratus | smooth layer of gray; may not be seen because of rain or snow | water droplets | 6,000 to 20,000 ft. or lower | widespread and continuous rain or snow | |
| **Low Clouds:** | | | | | |
| Stratus | smooth, even sheet | water droplets | below 5,000 ft. | drizzle | |
| Stratocumulus | more uneven than stratus; light and dark patches on underside | water droplets | below 5,000 ft. | overcast | follows storm |
| **Clouds Through All Levels:** | | | | | |
| Cumulus | heaped up in piles | water droplets | up to 20,000 ft. | | in morning, precedes storm; in afternoon, follows storm |
| Cumulonimbus | deep piles; may be anvil-shaped on top (cirrus layer) | water droplets; maybe ice crystals on top | usually 5,000 to 16,000 ft. thick; can reach 60,000 ft. thick | thunderstorms; heavy rain; hail possible | in winter, followed by north wind and colder weather |

*The Weather Report © 1989*

CLOUDS AND FOG    39

NAME _____

# NAME THAT CLOUD

Using what you have learned about cloud shapes and altitudes, label the clouds in the chart below.

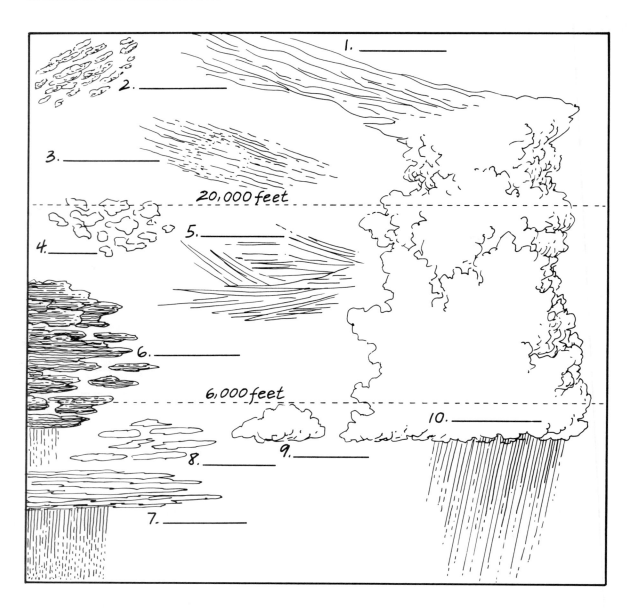

## Cloud Names

Altocumulus (Ac)    Cirrocumulus (Cc)    Cirrus (Ci)    Cumulus (Cu)

Stratocumulus (Sc)    Altostratus (As)    Cirrostratus (Cs)

Cumulonimbus (Cb)    Nimbostratus (Ns)    Stratus (St)

CLOUDS AND FOG

NAME _____

# Clouds and Fog Review Test

1. Draw a line between each type of cloud and the type of weather it will produce.

   Nimbus                Fair weather

   Cirrus                Thundershowers

   Cumulus               Rain

   Cumulonimbus          Fair, with rain possible within two days

2. Label the following parts of the water cycle model:

   - Prevailing winds
   - Clouds
   - Rain
   - Source of water vapor

3. In each blank below, write the level at which the cloud is found: high, middle, or low.

   Stratus _____

   Cirrus _____

   Cirrocumulus _____

   Cumulus _____

   Altostratus _____

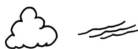

CLOUDS AND FOG   41

# CHAPTER 5
# PRECIPITATION

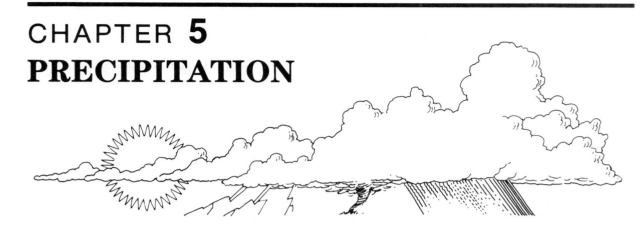

Precipitation is moisture that falls from clouds. Rain, snow, sleet, and hail are forms of precipitation.

The water vapor attaching to nuclei in clouds becomes precipitation when the water droplets become too heavy to be held aloft by air currents. Water droplets become heavier by bumping into each other and sticking together, or by colliding with ice crystals and then freezing and sticking together. This combining of water vapor droplets is called coalescing.

## *Did you know that . . .*

a cloud can develop and produce rain in less than an hour?

## Concept

Droplets of water vapor can coalesce into heavier drops and fall to earth.

## Activities

1. **Making rain.** Put a tray of ice over a boiling steam kettle that is giving off steam. Turn off the room lights and hold a flashlight under the kettle. A cloud should form and rain will fall. The cooling of the air decreases its ability to hold water and it becomes saturated.

2. **Measuring raindrops.** During a rain shower, briefly put a pan of flour outside. After a few drops have hit the flour, bring the pan inside and measure the size of the drops. Repeat this experiment on other rainy days and compare the size of the raindrops. The drops of drizzle from stratus clouds will be smaller than the drops during showers from cumulus or cumulonimbus clouds, because they have fallen from a lower altitude and therefore have encountered fewer water droplets to coalesce with. Drops from one cloud may vary too, because of varying temperatures and other conditions within the cloud. Drops that have originated as ice crystals are likely to be larger than drops that originated as water.

43

# Did you know that...

it takes an average of nine minutes for a raindrop to reach the ground from a cloud 20,000 feet thick?

## Concept

The form that precipitation takes—rain, snow, sleet, or hail—depends on the temperature of the various layers of air it passes through on the way to the ground.

## Activities

1. **Forms of precipitation.** Use the page "Forms of Precipitation" in an opaque projector, or duplicate it and hand it out to your students to clarify the varieties of precipitation.

    Rain can originate either as water or as ice. If sleet or snow passes through a layer of warm air, it will melt and reach the earth as rain.

    Sleet consists of round, hard balls of ice about the size of raindrops. It is formed when rain or half-melted snow falls through a layer of below-freezing air and freezes. (Very cold rain freezes when it touches something that is below freezing, such as the ground, pockets of air, or objects.) Because sleet can readily melt if it passes through a layer of warm air, it occurs only in winter.

    Hail is formed by the collision of supercooled water droplets with small ice crystals. It is formed when there are strong updrafts that bounce the hail between moister and drier areas; every time it bounces, the hailstone adds another layer of ice, so the bigger the hailstone, the longer it has been bouncing around in the clouds. Pilots have spotted hail when they are several miles away from a cloud. Sometimes hail is shot out of the sides of clouds because of the force and instability of the air currents. To reach golf-ball size, a hailstone must remain in the cloud for at least 20 minutes and must be bounced by the updraft at least two or three times before falling to earth. Hail is most common during spring and summer, usually during thundershowers.

    Snow falls to earth either as individual crystals or as snowflakes, which are large masses of crystals. In the atmosphere, snow grows directly from water vapor into solid crystals, formed around microscopic particles of dust or ice. Snow crystals that fall through very cold, dry air will not accumulate many other crystals and will fall as dry snow, which has a very low water content; as much as 20 inches of dry snow would have to melt to make an inch of water. Snow crystals that fall through moist air that is not too cold grow much larger and fall as wet snow; six or seven inches of wet snow can make an inch of water.

2. **Reading a weather map: precipitation.** Hand out the worksheet "Mapping Precipitation." This is the weather map from Chapter 2, showing the high and low temperatures in selected cities. Students will now add precipitation patterns to the map.

44   THE WEATHER REPORT

ANSWER KEY:
**Mapping Precipitation**

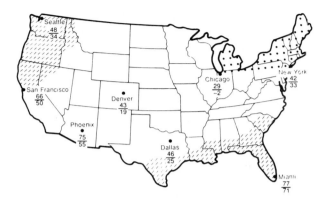

## Concept

Snow crystals usually form in hexagonal, or six-sided, shapes—flat plates, branching forms, or six-sided needles.

## Activities

1. **Snowflake shapes.** If you live in an area of the country where there is snow in the winter, you can show your students the symmetry and variety of snowflakes by going outside and catching some flakes on a dark piece of cloth. Examine the snowflakes with a magnifying glass.

    If you live in a more temperate part of the country, show your students the page "Snowflake Shapes" on an opaque projector to demonstrate this concept.

# *Did you know that . . .*

the largest hailstone on record measured over 5.5 inches in diameter! It fell on Coffeyville, Kansas, on September 3, 1970.

PRECIPITATION 45

## Concept

Supercooled water will instantly freeze when it touches ice. This is how hail begins.

## Activities

**From supercooled water to ice.** In a glass beaker, put ice water and salt. (Salt will bring the temperature well below freezing.) Put a test tube filled with water into the middle of the beaker. Wait about ten minutes for the water in the test tube to become supercooled. Lift the test tube slowly out of the beaker. Drop a tiny piece of ice into the test tube. What happens? The water will instantly freeze.

Cloud droplets frequently exist in a supercooled state because of conditions made possible in the higher atmosphere. On earth, water can be supercooled only by artificial means.

## Concept

Rain can be measured.

## Activities

1. **Coffee can rain gauge.** Put an empty coffee can out in the rain and measure the accumulated water. One inch in the can will equal one inch of rain. Compare it with a real rain gauge's reading.

2. **Reading a rain gauge.** To give your students practice reading a rain gauge marked off in tenths of inches, use the page "Reading a Rain Gauge" in an opaque projector. Cut out the water-level strips and put them on the gauge.

   Once the students have mastered reading by tenths and "reading between the lines," hand out the worksheet "How Much Rain?" Have them work through both sets of questions, and then go through the answers with the whole class.

3. **Comparing average and actual rainfall.** Put three poster-size rain gauges on your bulletin board at the beginning of the year. Use one to keep track of the actual rainfall. Use the second to show the average rainfall rate for your area. Use the third to show the rainfall for the preceding year. Bring the second and third gauges up to date when appropriate and compare the three totals. Have you had more or less rain than last year? Is the rainfall above or below average?

4. **Monthly rainfall average.** Graph the monthly rainfall averages for your city on a large graph on the bulletin board, or duplicate graphs for the students so they can each calculate the monthly rainfall average. (For rainfall data, see the Appendix or contact the nearest National Weather Service station.)

ANSWER KEY:
**How Much Rain?**

**A.** (Answers for amounts between marked increments may vary slightly.) **1.** 0.05 inch. **2.** 0.35 inch. **3.** 0.72 inch. **4.** 0.98 inch. **5.** 1.42 inches.
**B.**

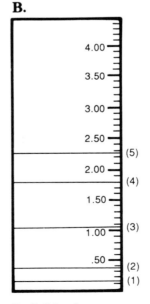

**6.** 5.6 inches.

46 THE WEATHER REPORT

## Concept

A rainbow is caused by sunlight shining through droplets of water, which act as prisms.

## Activities

1. **Making rainbows.** Use a prism or a crystal to show your class how white light is broken up into a spectrum by a prism.

   Now move outside to make a rainbow with a hose (using a fine spray) or a water spray bottle. The sun, the eye of the observer, and the water must all be in a straight line, with the sun behind the observer and less than 42° above the horizon, for the rainbow to be seen. You can experiment with different positions to show the alignment necessary to see a rainbow.

2. **Rainbow as weather predictor.** Use the page "Rainbows and Rain" to illustrate how the position of a rainbow indicates the location of rain. Because the sun cannot be high for a rainbow to be visible, rainbows are seen only in the early morning and in the late afternoon. A morning rainbow is usually in the west; then it is an indicator of rain later in the day, because weather generally moves from west to east. An afternoon rainbow, which is usually seen in the east, means that the rain is to the east—in other words, that it has gone past the observer.

PRECIPITATION

ANSWER KEY:
**Precipitation Review Test**

**1.** dew.  **2.** Las Vegas: fair. Portland, Oregon: rain. Boston: snow.  **3.** (Answers may vary slightly.) **A.** .21 inches. **B.** .72 inches. **C.** 1.29 inches. Total: 2.22 inches.

**4.** (Diagram should look something like this, showing cyclical wind patterns within the cloud that carry droplets upward, where they freeze and then fall—a cycle that is repeated several times before the piece of hail falls from the cloud.)

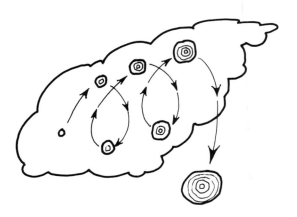

NAME _____

# FORMS OF PRECIPITATION

All forms of precipitation start as water vapor that condenses around microscopic particles.

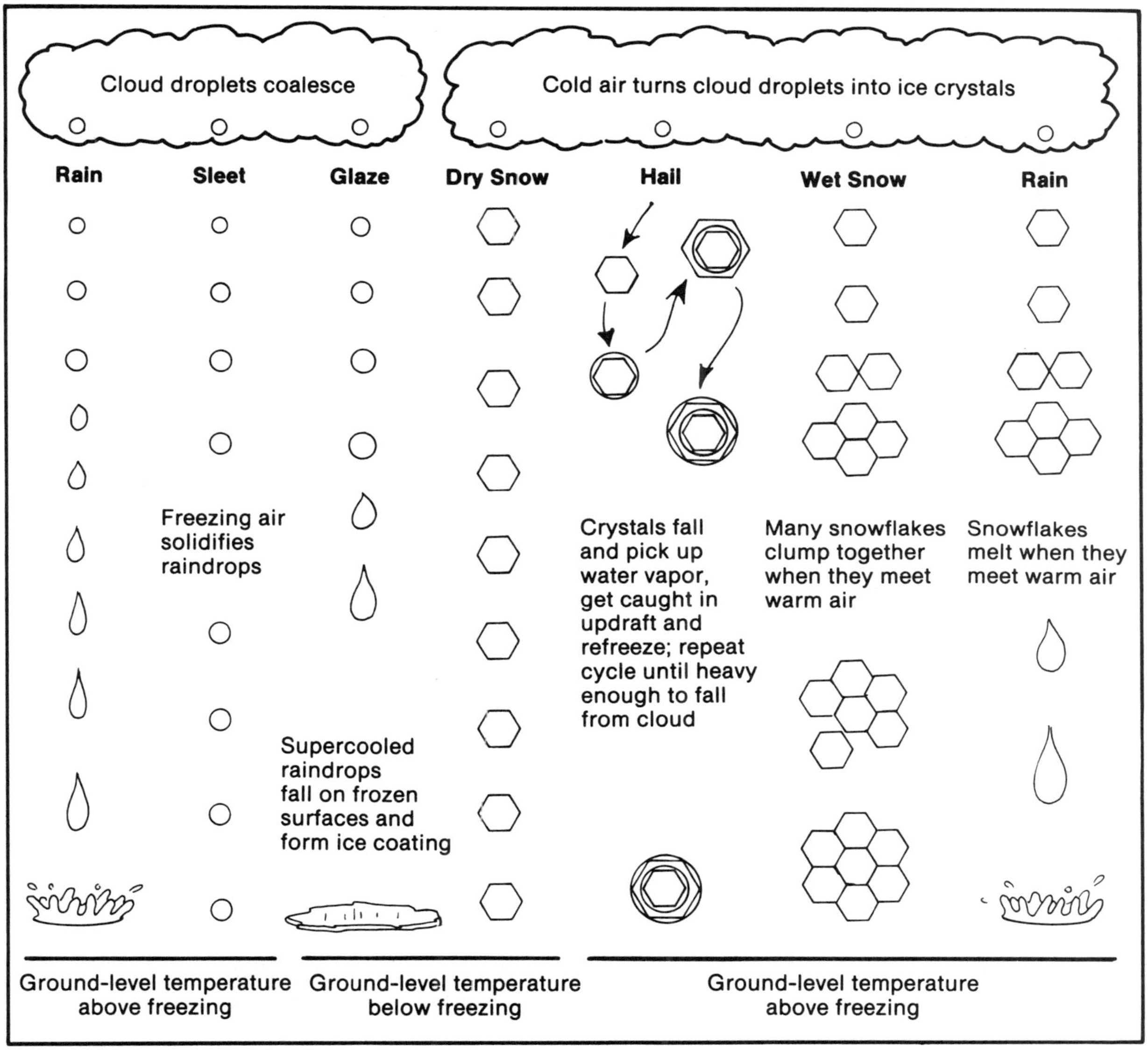

Cloud droplets coalesce

Cold air turns cloud droplets into ice crystals

**Rain**   **Sleet**   **Glaze**   **Dry Snow**   **Hail**   **Wet Snow**   **Rain**

Freezing air solidifies raindrops

Supercooled raindrops fall on frozen surfaces and form ice coating

Crystals fall and pick up water vapor, get caught in updraft and refreeze; repeat cycle until heavy enough to fall from cloud

Many snowflakes clump together when they meet warm air

Snowflakes melt when they meet warm air

Ground-level temperature above freezing

Ground-level temperature below freezing

Ground-level temperature above freezing

*The Weather Report* © 1989

PRECIPITATION     49

NAME _____

# MAPPING PRECIPITATION

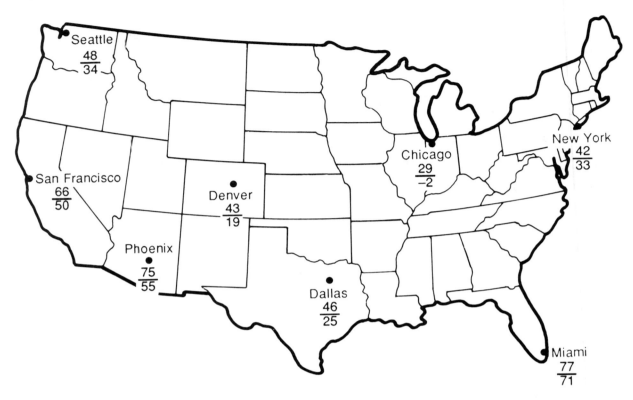

Use these symbols to show precipitation on the weather map:

Record these areas of precipitation on the weather map.

1. It's raining on the West Coast north of San Francisco, through Oregon, the western corner of Nevada, Washington, and the northern tip of Idaho.

2. It's snowing all throughout the northeast, from Chicago to New York and northward.

3. It's raining in the Gulf states—Florida, most of Georgia and South Carolina and the southern parts of Alabama, Mississippi, Louisiana, and Texas.

# SNOWFLAKE SHAPES

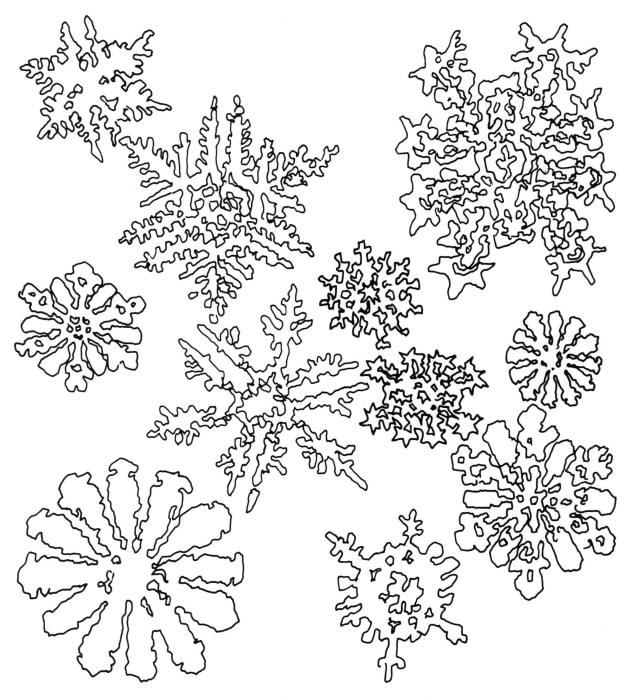

It can snow at 36° or even warmer. Whether the precipitation reaches the ground as snow or turns to rain on the way down depends on the downdraft in the cloud and on the wind. The wind evaporates the moisture in the air and keeps the snowflakes cooler than the air. Much rainfall begins as snow but melts on the way down, usually around 1,000 feet below the freezing level.

# READING A RAIN GAUGE

A rain gauge is marked in tenths of inches.

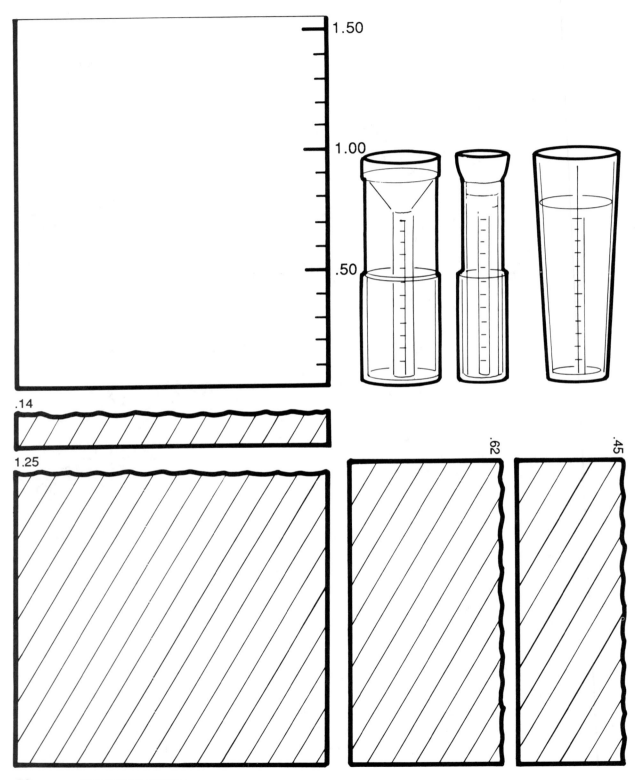

NAME _____

# HOW MUCH RAIN?

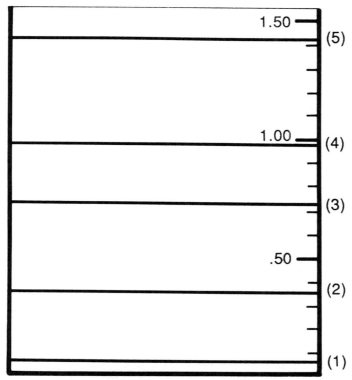

*(enlarged size)*

A. Write the amount of rain at each line.

1. _____
2. _____
3. _____
4. _____
5. _____

B. Draw lines on this actual-size rain gauge to indicate the following amounts of rain. Number each line.

1. .13 inches
2. .36 inches
3. 1.05 inches
4. 1.80 inches
5. 2.26 inches
6. What is the total rain for these 5 days?

   _____

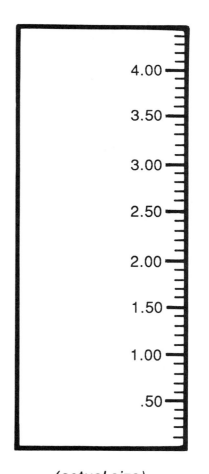

*(actual size)*

PRECIPITATION 53

# RAINBOWS AND RAIN

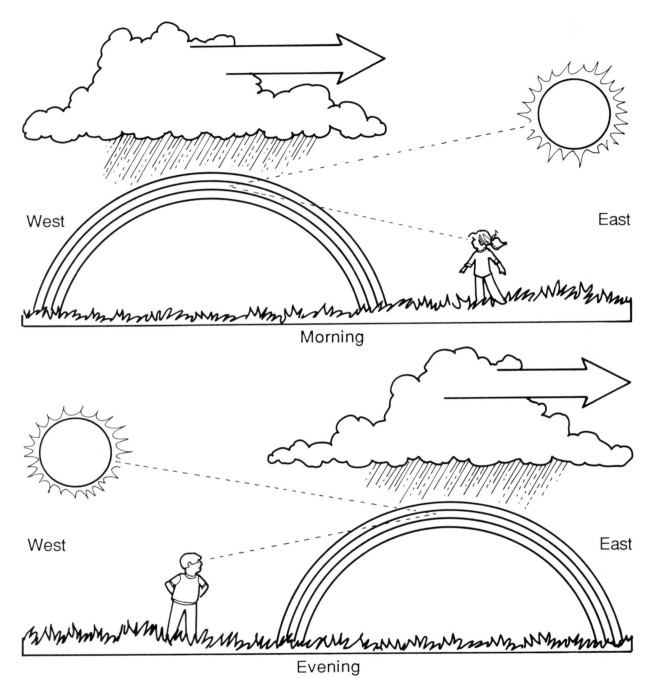

To see a rainbow, the sun must be at your back.

A rainbow in the morning means that rain may be coming your way.

A rainbow in the afternoon means that the rain is past.

NAME _____

# Precipitation Review Test

1. Circle the one that is not a form of precipitation.

    rain        dew        hail        snow

2. What kind of weather does the map show for the following cities:

    Las Vegas _____

    Portland _____

    Boston _____

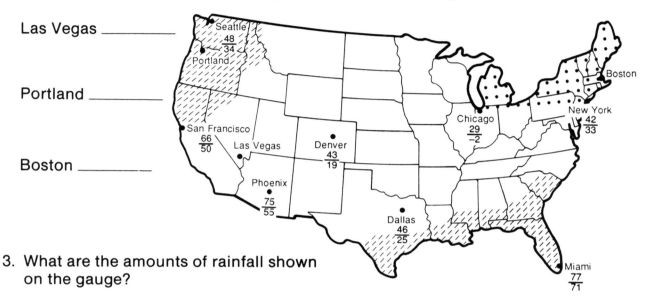

3. What are the amounts of rainfall shown on the gauge?

    A. _____

    B. _____

    C. _____

    What is the total of these three rainfalls?

    _____

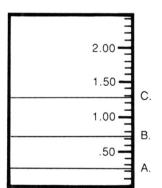

4. In this cumulonimbus cloud, make a diagram showing the process of hail formation.

PRECIPITATION   55

# CHAPTER 6
# WIND

Wind is the flow of air over the earth's surface. It is primarily caused by the uneven heating of the earth's surface by the sun. Warm air expands, becomes lighter, and rises. Cooler air rushes in to fill the space left by the rising warm air. The earth's rotation deflects the wind, steering it in various directions.

## Concept

Warm air expands and rises. Cool air sinks.

## Activities

1. **Warm air expands and rises.** Put a balloon over the mouth of a soda bottle. Put the bottle in a bowl of very hot water. What happens? Now put the bottle in a bowl of very cool water. What happens?

2. **Warm air rises (I).** Cut out a spiral from aluminum foil. Attach it to a piece of thread so that it can spin. Hold the spiral so the bottom end is about 12 inches above a hot plate. Turn the hot plate on. Watch the rising hot air make the spiral spin.

3. **Warm air rises (II).** Chill an empty soda bottle in the freezer until it becomes very cold. Put a small coin on the mouth of the bottle. Rub your hands along the bottle to warm the air inside, or put the cold bottle in warm water. Watch the coin. As the air warms, it expands and rises. The rising, expanding air pushes against the coin, causing it to lift up. Once the air is released, the coin will again cover the mouth of the bottle, but if the bottle is continuously rubbed, this action will be repeated.

4. **Warm air rises and cool air sinks (I).** Dip the open end of a small empty can in bubble soap so it is covered with a layer of soap film. Set the can in 1 inch of hot water. Watch the soap film expand. Now set the can in 1 inch of cold water. Watch the soap film contract.

5. **Warm air rises and cool air sinks (II).** Attach a paper bag, open end down, to each end of a yardstick. Tie a string to the middle of the yardstick, which you will use to suspend it. Hold a candle under one of the bags, being careful not to let it burn. What happens? The warmed bag of air will become lighter and rise.
   Next, take an empty jar out of the freezer and immediately hold it over, or near, one of the bags. What happens? The cooled bag of air will sink.

## Concept

Wind is caused by a temperature difference, with warm air, which is light, being pushed up by cool air, which is heavier. This movement is called convection.

# Activities

1. **Convection current.** With this demonstration you can show the movement of warm air being pushed up by cooler air.

   In one of the long sides of a shoe box, cut out a rectangular viewing hole and cover it with plastic wrap. Cut a 1½-inch diameter hole toward each end of the bottom of the shoe box. Make a chimney for each hole by taping a 3" × 5" index card into a circle (with a circumference of about 4½ inches, to fit the hole). Tape each chimney securely over a hole.

   Light a small candle. Carefully place the box, open end down, over the candle, so that one of the chimneys is directly above the candle. Hold a smoking (not flaming), tightly rolled paper towel over the other chimney. The smoke will be drawn into that chimney and come out of the other chimney. The warm air above the candle is being pushed up by cooler air, causing a convection, and the cool air near the other chimney is moving over to replace the rising warm air.

2. **A visible convection.** Convection occurs in liquid as well as in air. This demonstration, using colored water, clearly shows the movement of convection.

   Make a medium-size hole in a film container and weight it with a few pennies. Fill the container with hot colored water. Gently drop the container into a tank of cool water. What happens? Why?

   Now repeat the experiment with cold colored water dropped into a tank of hot water. What happens? Why?

   Heat makes water (like air) expand and rise. Cool water (like air) is heavier than warm water and will sink.

3. **Sea breeze.** This demonstration of convection uses a model of a coastal area to show the movement of air between water and land.

   Build a large rectangular box of wood or of lucite plastic (with solvent glue) at least 2½ feet long. Paint a board with flat black paint (this is the land) that is the width of the box and less than half as long. Place the board in one side of the box. In the middle of the other half,

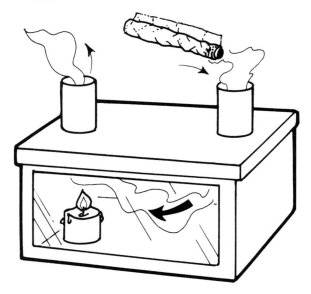

58   THE WEATHER REPORT

securely place a candle (or a smoking match on a jar lid). Position a floodlight or a heat lamp (this is the sun) above the board so it will shine on it. Carefully pour a 1-inch layer of ice water (the sea) into the box.

Turn on the lamp. Light the candle and then blow it out. Watch where the smoke goes. Here's what your box should look like.

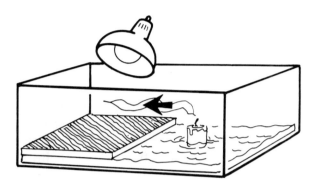

Cool air replaces rising hot air. Because land heats more quickly than water (as was demonstrated in the chapter on temperature), this experiment shows how air moves during the daytime over cities with large bodies of water to the west of them. So the smoke will move to the board (the land) and then rise. The opposite (although not as obviously) happens at night as the land loses heat faster than water.

## Concept

Dust devils, the small, dry swirls of wind often seen in deserts, are the result of very small-scale convections.

## Activity

**Dust devil model.** Cut a large coffee can down one side so there is a ½-inch to 1-inch opening. Build a little paper fire in the can. Watch the wind pattern rise. When the heat rises, the cool air from the crack replaces the warm air, causing the swirling patterns.

## Concept

When a layer of cold air is trapped beneath a layer of warm air, an inversion results, and smog and other airborne pollution is trapped until wind makes the air circulate again.

## Activity

**Making an inversion.** This demonstration requires an aquarium, blue ink, and a funnel with a tube.

Fill the aquarium one-third full with cold water and ice. Next, gently pour in boiling water (out of the kettle) until the aquarium is two-thirds full. Because you want to avoid mixing the cold and the hot water, you should slowly pour the boiling water onto a board floating on top of the ice water. Mix the ink with warm water and pour it into the funnel attached to a tube that runs into the bottom of the aquarium. What happens?

The blue liquid is warmer than the cold water at the bottom, so the blue liquid will rise. Because the blue liquid is cooler than the hot water, however, it can rise only to the top of the cold water. The blue liquid represents smog, and it is trapped—just as smog becomes trapped over Los Angeles and Denver by this kind of inversion of the air. Tule fog in California's San Joaquin Valley also becomes trapped in this way.

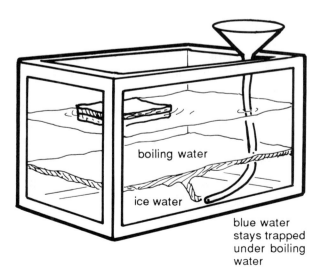

WIND 59

## Concept

Wind variations, or eddies, are caused by difference in surface and terrain.

## Activities

1. **Down drift.** Draw the cardinal directions on a 12″ × 18″ piece of paper. Put this paper in different parts of the room (always being sure that North on the paper is pointing north), and drop a piece of down or a small bit of cotton over the center of it. Does the feather always land in the same direction? Why or why not? What would happen in a open field? What would happen near trees and buildings?

2. **Are trees shelter from the wind?** Put a candle behind a soda bottle. Blow out the candle from the other side of the bottle. What happens? The bottle will make it harder to blow out the candle.

   Go outside with a wind-speed instrument. Measure the wind speed out in the open. Measure the wind speed behind a tree. Compare the results. The tree blocks the wind somewhat, but the wind still goes around the tree.

## Concept

Wind direction can be observed.

## Activities

1. **Look at the flag.** If there is enough wind to stir the flag at your school, it will be pointing in the direction toward which the wind is blowing. (Remember that winds are identified by the direction that they are blowing from, not by the direction they are blowing toward.)

2. **Use your finger.** Wet your finger and stick it up in the wind. The side that evaporates first is the direction that the wind is coming from.

3. **Making a wind vane.** See page 141 in the chapter "Building Your Own Weather Station."

4. **Making a windsock.** See page 142 in the chapter "Building Your Own Weather Station." (Note that a windsock is used by pilots to observe both the force and direction of the wind.)

## Concept

Wind speed can be observed.

## Activities

1. **Beaufort Wind Scale.** Ask students to name ways that they can tell the wind is blowing (for example, hair blows, leaves rustle, flags flutter). Then hand out the reproducible worksheet "How Windy Is It?" The Beaufort Wind Scale is a guide for estimating wind speed. Discuss the indicators of wind speed and the names for the various kinds of wind with your students. Have them observe the signs of wind speed for a week and record them on the chart on the worksheet. Discuss the results at the end of the week.

   Because winds tend to increase in the afternoon, due to vertical mixing of warm and cool air, students could also record their observations of the wind twice a day, in the morning and in the afternoon. At the end of the week, have them compare the morning readings with the afternoon readings.

2. **A wind scale for the bulletin board.** Divide your students into 13 groups. Have each group make a 12″ × 18″ picture illustrating a different wind speed from the Beaufort scale. Label each picture and put them up on a bulletin board.

## Concept

Wind speed can be measured.

## Activities

1. **Wind-speed card.** Hand out the reproducible worksheet "A Wind-Speed Meter." Help your students make their own wind-speed card. (Be sure to have enough index cards and paper clips on hand.) Let them experiment with wind speeds at various places on the school grounds.
2. **Making an anemometer.** See page 143 in the chapter "Building Your Own Weather Station."
3. An accurate, inexpensive, and durable instrument for measuring wind speed is the Dwyer Wind Meter, which is available through scientific supply catalogs.

## Concept

Wind speed increases with elevation because of reduced frictional drag.

## Activity

**Wind speed increases with elevation.** Measure the wind speed with an anemometer or wind meter at various elevations around the school (for example, 1 foot, 3 feet, 5 feet, and higher if possible). Compare the various speeds with the elevations.

## Concept

The wind chill is caused by wind blowing away warm air molecules around your body and making you feel colder. This effect is more pronounced when you are wet, because the evaporation of moisture takes heat energy away from your body and thus cools your body. Thus, hypothermia often occurs when people are wet and the temperature is between 30° and 50°—fairly moderate temperatures that people do not usually think are dangerous.

## Activity

**Wind chill.** Hand out the worksheet "The Wind-Chill Factor." After the students have answered the questions, have them calculate the wind-chill factor for the day (round off the actual temperature and the wind speed). Because the wind-chill factor can change during the day, especially during the winter, your students might take this measurement several times during the day. Discuss the effects of wind chill on the body and precautions to take against getting too cold.

---

## *Did you know that...*

the strongest wind ever recorded on earth was 230 miles per hour! It was recorded at the summit of Mount Washington, New Hampshire, (elevation 6,262 feet) on April 12, 1934.

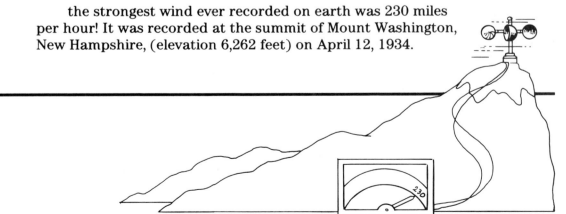

WIND 61

ANSWER KEY:
### The Wind-Chill Factor

**1.** -2°. **2.** 26°. **3.** 37°. **4.** Temperature = -40°; wind speed = 40 mph. **5.** Answers will vary.

## Concept

Because of the effects of the rotation of the earth, there are six belts of wind around the earth that form where the air heats and rises or cools and sinks. These belts determine the prevailing direction of the wind in any area.

## Activities

**Prevailing winds.** Duplicate page 66, "Global Wind Patterns," for your students. Use it to discuss the primary wind patterns.

If the earth did not rotate, the winds would flow in one continuous north-south cycle from the equator toward each pole, where they would become cold and sink back toward the equator. Instead, the air flows in six bands, or prevailing patterns, because it travels at different speeds depending on its latitude.

Have the students identify the prevailing wind where you live.

---

## *Did you know . . .*

how the horse latitudes got their name? In the 1500s ships sailing from Spain to the Americas became becalmed in the still air here. In order to have enough drinking water for the people on the ships, they had to throw the horses overboard.

---

ANSWER KEY:
### Wind Review Test

**1. A.** Daytime    **B.** Nighttime

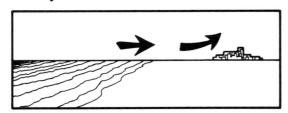 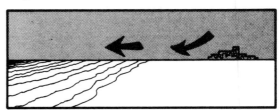

**2.** False.  **3.** True.  **4.** True.
**5.** True.  **6.** False.  **7.** True.

**62    THE WEATHER REPORT**

NAME _____

# HOW WINDY IS IT?

You can estimate the speed of the wind by watching how it affects things outside, like trees, flags, and chimney smoke. Below is the **Beaufort Wind Scale**, which you can use to match the effects of the wind to its speed. Observe the wind's effects at the same time of day for a week or at various predetermined times, and record your observations and your estimate of the wind speed in the chart at the bottom of the page.

| Observation | Name of Wind | Miles per Hour | Symbol |
|---|---|---|---|
| Smoke goes straight up | Calm | Less than 1 | |
| Smoke moves, but wind vane does not | Light air | 1–3 | |
| Leaves rustle, wind vane moves, wind felt on face | Light breeze | 4–7 | |
| Leaves and small twigs move constantly, wind extends light flag | Gentle breeze | 8–12 | |
| Dust raised, dead leaves and loose paper blows about, small branches move | Moderate breeze | 13–18 | |
| Small trees sway, small waves crest on lakes or streams | Fresh breeze | 19–24 | |
| Large branches move constantly, wind howls around eaves, wires on telephone poles hum | Strong breeze | 25–31 | |
| Large trees sway, walking against wind is inconvenient | Moderate gale | 28–32 | |
| Twigs break off trees, walking against wind is difficult | Fresh gale | 39–46 | |
| Branches break off trees, loose bricks blown off chimneys, shingles blown off | Strong gale | 47–54 | |
| Trees snap or are uprooted, considerable damage to buildings is possible | Whole gale | 55–63 | |
| Widespread damage to buildings | Storm | 64–75 | |
| General destruction | Hurricane | Over 75 | |

| | Observations | Name of Wind | Miles per Hour | Symbol |
|---|---|---|---|---|
| Monday | | | | |
| Tuesday | | | | |
| Wednesday | | | | |
| Thursday | | | | |
| Friday | | | | |

*The Weather Report* © 1989

WIND 63

NAME _____

# A WIND-SPEED METER

You can measure wind speed with an index card and a paper clip. Here's how.

1. Cut out the wind-speed meter pattern.//
2. Glue it to a 3" × 5" index card. Be sure to put glue all over the back of the pattern.
3. When the glue is dry, cut along the line that says "cut."
4. Fold along the two lines that say "fold."
5. Put a paper clip at the end of the cut strip. Adjust the folds so it lines up with the "0". Your wind-speed meter should look like this.

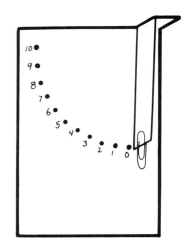

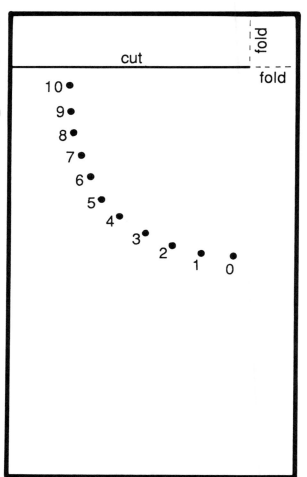

Now you're ready to use your windspeed meter. Here's how.

1. Hold it up facing the wind. The wind will move the paper-clip pointer.
2. Read the number by the pointer. That's the speed of the wind. If the pointer keeps moving between two numbers, the wind speed is in that range. For example, if the pointer keeps moving between 3 and 5, the wind speed is 3-5 miles per hour.
3. Use your wind-speed meter to measure wind speed in various places.

64   WIND

NAME _____

# THE WIND-CHILL FACTOR

| Wind Speed (mph) | Thermometer Readings (°F) | | | | | | | | | |
|---|---|---|---|---|---|---|---|---|---|---|
| | 50 | 40 | 30 | 20 | 10 | 0 | -10 | -20 | -30 | -40 |
| | Equivalent Temperatures (°F) | | | | | | | | | |
| Calm | 50 | 40 | 30 | 20 | 10 | 0 | -10 | -20 | -30 | -40 |
| 5 | 48 | 37 | 27 | 16 | 6 | -5 | -15 | -26 | -36 | -47 |
| 10 | 40 | 28 | 16 | 4 | -9 | -21 | -33 | -46 | -58 | -70 |
| 15 | 36 | 22 | 9 | -5 | -18 | -36 | -45 | -58 | -72 | -85 |
| 20 | 32 | 18 | 4 | -10 | -25 | -39 | -53 | -67 | -82 | -96 |
| 25 | 30 | 16 | 0 | -15 | -29 | -44 | -59 | -74 | -88 | -104 |
| 30 | 28 | 13 | -2 | -18 | -33 | -48 | -63 | -79 | -94 | -109 |
| 35 | 27 | 11 | -4 | -20 | -35 | -49 | -67 | -82 | -98 | -113 |
| 40 | 26 | 10 | -6 | -21 | -37 | -53 | -69 | -85 | -100 | -116 |
| | little danger | | | increasing danger | | | great danger | | | |

1. How cold does it feel when the wind speed is 30 mph and the temperature is 30°? _____

2. How cold does it feel when the wind speed is 40 mph and the temperature is 50°? _____

3. How cold does it feel when the wind speed is 5 mph and the temperature is 40°? _____

4. What are the actual temperature and the wind speed when it feels like -116°? _____

5. What is the actual temperature right now? _____

   What is the wind speed right now? _____

   How cold does it feel? _____

The Weather Report © 1989

WIND 65

# GLOBAL WIND PATTERNS

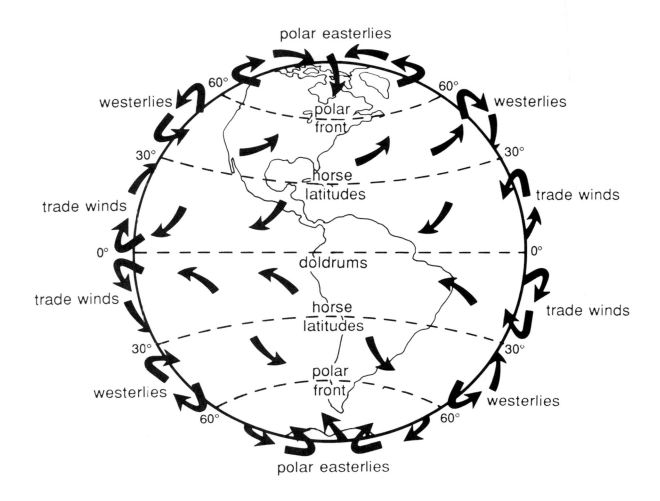

The doldrums are near the equator. The air is calm here because there is little change in temperature.

Above the equator the trade winds are steady northeast winds. Below the equator, the trade winds are steady southeast winds.

The horse latitudes are another area of calm. The air is cooling and sinking here.

The westerlies are the prevailing wind over much of the middle and higher latitudes.

The polar front is where the warm air of the westerlies meets the cold air of the polar easterlies. This causes unstable weather in the region where the westerlies prevail.

NAME _____

# Wind Review Test

1. Draw arrows in the pictures below to show the direction of the wind.

   A. Daytime

   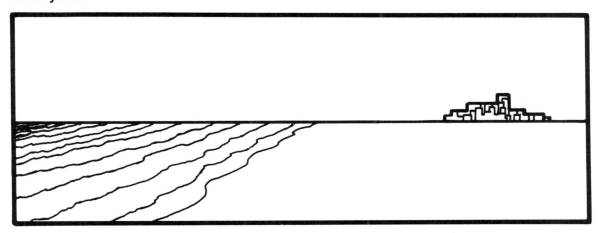

   B. Nighttime

   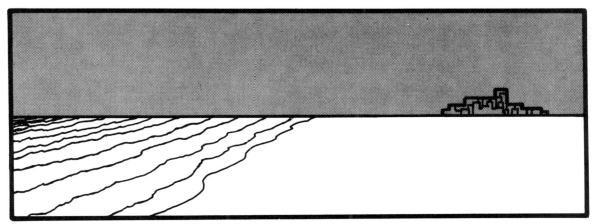

2. True _____ or False _____ .   Water heats faster than land.

3. True _____ or False _____ .   Cool air can be trapped under warm air during an inversion.

4. True _____ or False _____ .   Trees can partially block the wind.

5. True _____ or False _____ .   Winds blowing toward the east are from the west.

6. True _____ or False _____ .   The windier it is, the warmer it feels.

7. True _____ or False _____ .   The wind-chill factor makes you feel cooler when you are wet.

# CHAPTER 7
# AIR PRESSURE

Because air has weight, it can exert pressure. The weight of the atmosphere, or the air over the whole earth, is constant, but it changes locally. The weight of the air over a given spot is called air pressure. Air pressure can be measured.

Warm, moist air rises and is associated with low, or weaker, pressure areas. As this air rises, it cools and the water vapor in it condenses; this is how clouds are formed. Cool air sinks and compresses, causing high pressure.

Rapidly falling air pressure means that a storm is coming. Rapidly rising air pressure means fair weather. Differences in terrain, and in individual storms can cause exceptions to this general rule, however.

Pressure systems are whirling masses of air, called highs or lows, that cover very large areas. Low-pressure systems usually bring cloudy skies and often rain or snow. High-pressure systems bring dry, gradually clearing weather or generally fair weather.

## Concept

The atmosphere, or all of the air around the earth, has weight and thus creates pressure. The weight of air over the whole earth is constant, but it changes locally.

## Activities

1. **The atmosphere.** Use page 74, "The Atmosphere," on an opaque projector to show your students how the atmosphere surrounds the earth.

## Concept

Air has weight and can exert pressure.

## Activities

1. **Air pressure (I).** For this demonstration, you'll need a large metal can with a screw-on top (such as the 1-gallon cans that hold duplicating fluid—make sure the can is thoroughly cleaned). Heat a ½ inch of water in the can to boiling. When you are sure the can is full of steam from the boiling water, turn the heat off and quickly screw the cap on very tightly. Put the can in a tray of cold water. What happens?

The can is full of steam, or water vapor with very little air left in it. When the lid is put on and the heat turned off, the water vapor condenses back to liquid water and thus decreases its volume by a factor of about 1,000. This huge decrease in volume causes the pressure to drop drastically inside the can. The greater pressure outside the can then crushes it.

This demonstration can also be done with a soda-pop can. Heat a ¼ inch of water in the can to boiling on a hot plate. Hold the can with a hot pad and put it upside down into a pan filled with 3 or 4 inches of cold water. The water will plug the hole and the can will collapse.

# Did you know that...

all the air around the earth weighs close to 5,600 trillion tons?

2. **Air pressure (II).** Tie an inflated balloon at each end of a ruler. Pop one of the balloons. What happens? The ruler will go up on the side of the popped balloon, showing that air has weight.

3. **Air pressure (III).** Fill a glass to overflowing with water. Lay a post card or an index card on top. Hold the card with one hand, while you turn the glass over. Remove the hand holding the card. What happens? The card will stay in place because the air pressure outside the glass is greater than the water pressure inside.

4. **Air pressure (IV).** Hold a soda bottle horizontally and put a small styrofoam ball just inside its neck. Blow into the bottle, with the ball just out of the way. The ball will shoot out due to the greater air pressure inside the bottle caused by blowing into it.

5. **Air pressure (V).** Crumple a napkin and put it at the bottom of a glass. Put the glass straight down into a tank of water. The napkin will stay dry due to the air pressure being greater than the water pressure.

6. **Air pressure (VI).** Tape one end of a straw into a sandwich bag. Set the bag under a book and blow into the straw. The book will rise due to the air pressure.

7. **Air pressure (VII).** Place about 2 inches of a long piece of paper into the pages of a book, with the rest of the paper hanging out lengthwise. With the loose end pointed away from you, blow along the top of the paper. It will temporarily rise because the air's pressure is momentarily displaced.

You can also do this experiment simply by blowing over a piece of paper held in both hands. Hold the paper near your lips.

## Concept

Air pressure can be measured.

## Activities

1. **Soap-bottle barometer.** See page 138 in the chapter "Building Your Own Weather Station."

2. **Tin-can barometer.** See page 139 in the chapter "Building Your Own Weather Station."

3. **Reading a barometer: demonstration.** Use page 75, "Reading a Barometer," on an opaque projector. Cut out the needle and use it to show various air-pressure readings as they would appear on an aneroid barometer. Note that although barometers commonly show a range of 26 or 28 to 31, the needle rarely moves to the extremes on either end. The air pressure is usually between 29.20 and 30.40, with 29.92 the average sea-level air pressure.

   Introduce the concept that air pressure is related to weather forecasting. The farther the pressure drops below the average 29.92, the greater the likelihood of a storm. The farther the pressure rises above 29.92, the greater the likelihood of fair weather.

   The numbers on the barometer represent numbers of inches of mercury in a glass tube (this was the first barometer, and modern versions of it are still in use). The air pressure holds the mercury up in the tube. When air pressure falls, it cannot hold the mercury as high. When air pressure rises, it pushes up the mercury.

4. **Reading a barometer: practice.** Give the students the worksheet "How Heavy Is the Air Today?" Have them answer the questions.

## *Did you know that . . .*

the lowest barometer reading ever recorded was 25.70—in a typhoon in the Philippine Sea on October 12, 1979. The highest barometer reading ever recorded was 32.01 on December 31, 1968, in Agata, Siberia.

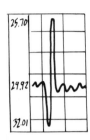

ANSWER KEY:
**How Heavy Is the Air Today?**

A.  **1.** 26.20.  **2.** 29.20.  **3.** 29.92.  **4.** 30.25.
    **5.** 31.60.

B.

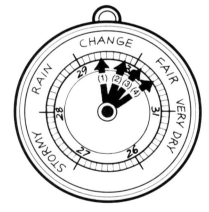

   **5.** Between 28 and 31.

AIR PRESSURE 71

## Concept

Air pressure can be used to predict weather.

## Activities

1. **Predicting weather with air pressure and wind direction.** Hand out the page "Predicting the Weather with Wind and Air Pressure." You might also make a poster-size version of this table and post it on the bulletin board for easy reference. Have the students answer the questions on the worksheet "What's Your Prediction?" so they will become familiar with the information contained in the table. You might also ask them to predict the weather in your area from the current wind and air pressure readings.

   Note that "Predicting the Weather with Wind and Air Pressure" is a general table; the terrain in your area may cause some variations. Be sure that your barometer is adjusted to sea-level pressure (check with your local weather station or airport) so it will correspond with the pressures given in the table. Barometric pressure varies with altitude: an increase of 1,000 feet in altitude causes a drop in air pressure equal to 1 inch of mercury.

2. **Air pressure during a storm.** To demonstrate the variations in air pressure during a rainstorm, use the chart, "Storm Passage in Van Nuys, California (1987)," on an opaque projector. Discuss how the changes in air pressure correspond to the weather changes.

   Note that wind direction during storm passage depends on where you are in the country. If you have observed typical changes in wind direction as a storm passes, you can use these patterns to tell when a storm front has passed. For example, if during a storm passage in your area, the winds usually shift from southwest to northwest, the clouds begin to break up and the air pressure begins to rise, then the shift of winds from southwest to northwest will generally be an indicator of frontal passage in your part of the country.

3. Chart a storm. Duplicate the worksheet "A Storm's Air-Pressure Profile" and hand it out on a day when a storm is approaching (12–24 hours before the storm, if possible). Have your students record the barometer readings every hour (you'll have to provide the readings for the before- and after-school hours). Discuss the profile of the storm. When was it raining? When did it clear? Can you tell by looking at the chart? You might also compare this storm record with the profile of the storm on page 80.

## ANSWER KEY:
**What's Your Prediction?**

1. Rain within 24 hours. **2.** Increasing wind; rain in 12–18 hours. **3.** (June) Rain, with high wind, then clearing within 36 hours; (December) Rain or snow, with high wind, then clearing and colder within 36 hours. **4.** Clearing within a few hours. **5.** End of storm—clearing and colder. **6.** Fair weather. **7.** Rain may be on the way. **8.** It's raining. **9.** Severe storms. **10.** Falling rapidly.

## Concept

The movement of high- and low-pressure systems, caused by the rotation of the earth and by the terrestrial (or prevailing) winds, brings weather changes.

## Activity

**Pressure systems.** To help your students visualize the global patterns of high- and low-pressure systems and their effects on weather, duplicate "Summer Storm Chart" (page 81) and hand it out.

   The curving lines around the high- and

low-pressure cells on this modified weather map (a modified 500-millibar chart) show the general wind pattern in the upper atmosphere.

In the regions of the prevailing westerlies, high-pressure systems are moved along by the wind from west to east. Because air flows from high-pressure areas into low-pressure areas, high-pressure systems push low-pressure systems ahead of them. In the Northern Hemisphere, the air in high-pressure cells moves clockwise, and the air in low-pressure cells moves counter-clockwise. Polar fronts, where warm and cold air meet, are the result of high- and low-pressure systems.

Have your students locate the high-pressure and the low-pressure centers (marked with H and L on the map) and color them. High-pressure systems bring dry weather. Low-pressure systems bring cloudy skies and often rain or snow. Next, have your students locate and color the storm lines (▽▽▽▽). If this were a satellite photograph, bands of clouds would appear where these lines are.

Where does the map show rainy weather? Where is it fair? For example, Southern California is protected by a ridge of high-pressure—fair weather. Portland, Oregon, however, is right in the path of a low-pressure system; it probably had periods of showers when this map was made. What kind of weather was your city having?

## ANSWER KEY:
### Air Pressure Review Test

**1.** 29.98 and falling slowly: Possible rain. 30.06 and rising slowly: Fair. 29.80 and falling rapidly: Rain.  **2.** C. 29.92.
**3. A.** 29.50.  **B.** 29.96.  **C.** 30.23.
**4. A.** Rain likely.  **B.** Frontal passage.  **C.** Clearing.

# THE ATMOSPHERE

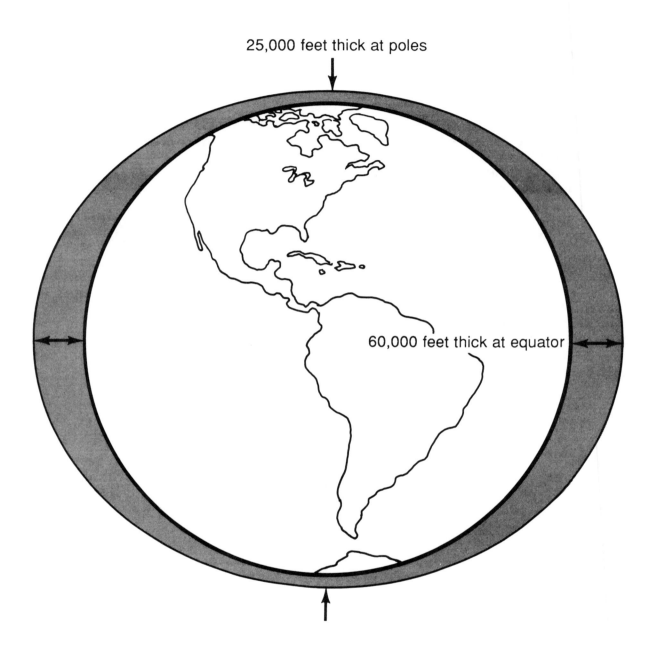

The atmosphere is all of the air around the earth. This blanket of air acts as protection from harmful rays of the sun and regulates the Earth's temperatures. Celestial bodies without atmospheres, such as the moon, are unprotected, so their days are extremely hot and their nights are extremely cold.

AIR PRESSURE

# READING A BAROMETER

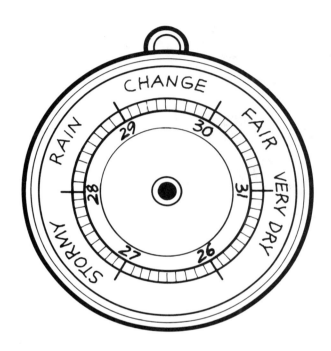

- - - - - - - - - - - - - - - - - - - - - - - - - - - - - - - - - - - - - - - - - - - -

AIR PRESSURE

NAME _____

# HOW HEAVY IS THE AIR TODAY?

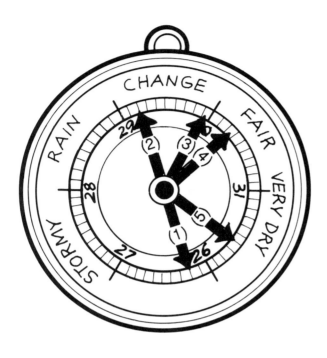

A. What barometer readings do the arrows show?

1. _____
2. _____
3. _____
4. _____
5. _____

B. Draw arrows on this barometer (and number them) to show the following readings:

1. 29.40
2. 29.80
3. 30.10
4. 30.40
5. Between which whole numbers will the needle most often be found?

   _____

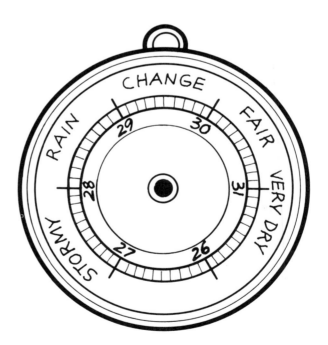

AIR PRESSURE

*The Weather Report © 1989*

# PREDICTING THE WEATHER WITH WIND AND AIR PRESSURE

| Wind Direction | Barometer Reading | General Weather Forecast |
|---|---|---|
| SW to NW | 30.10 and above—steady | Fair, with little temperature change. |
| SW to NW | 30.10 to 30.20—rising rapidly | Fair, followed within 2 days by rain. |
| SW to NW | 30.20 and above—falling slowly | Fair and slowly rising temperature for 2 days. |
| S to SE | 30.10 to 30.20—falling slowly | Rain within 24 hours. |
| S to SE | 30.10 to 30.20—falling rapidly | Increasing wind; rain in 12–24 hours. |
| SE to NE | 30.10 to 30.20—falling slowly | Increasing wind; rain in 12–18 hours. |
| SE to NE | 30.10 to 30.20—falling rapidly | Increasing wind; rain within 12 hours. |
| E to NE | 30.10 and above—falling slowly | Summer: rain in 2 to 4 days. Winter: rain or snow within 24 hours. |
| E to NE | 30.10 and above—falling rapidly | Summer: rain in 12–24 hours. Winter: rain or snow, with increasing winds, within 24 hours. |
| SE to NE | 30.00 or below—falling slowly | Rain will continue for 1 to 2 days. |
| SE to NE | 30.00 or below—falling rapidly | Rain (or snow) with high wind; then clearing (and colder in winter) within 36 hours. |
| S to SW | 30.00 or below—rising slowly | Clearing within a few hours. |
| S to E | 29.80 or below—falling rapidly | Severe storm within a few hours, followed within 24 hours by colder weather. |
| E to N | 29.80 or below—falling rapidly | Severe NE gale with heavy precipitation. In winter, heavy snow followed by cold wave. |
| Going to W | 29.80 or below—rising rapidly | End of storm—clearing and colder. |

NOTE: A rapid rise or fall is 0.05 to 0.09 inches or more in 3 hours; a slow rise or fall is less than 0.05 inches in 3 hours.

*The Weather Report* © 1989

NAME _____

# WHAT'S YOUR PREDICTION?

To answer the following questions, refer to the chart "Predicting the Weather with Wind and Air Pressure."

1. The wind is S to SE. The barometer is at 30.10 and has been falling slowly. What's your prediction?

   _____

2. The wind is SE to NE. The barometer is at 30.10 and has been falling slowly. What's your prediction?

   _____

3. The wind is SE to NE. The barometer is at 30.00 and is falling rapidly. What would you predict if this happened in June?

   _____

   What would you predict if this happened in December?

   _____

4. The wind is S to SW. The barometer is at 29.80 and has been rising rapidly. What's your prediction?

   _____

5. It is stormy. Now the wind is shifting to the west. The barometer is at 29.80 and is rising rapidly. What's your prediction?

   _____

6. In general, barometer readings of 30.10 and above that are steady or rising mean (circle the correct answer)

           fair weather        it's raining        severe storms

7. In general, barometer readings of 30.10 and above that are falling mean (circle the correct answer)

           rain may be on the way        it's raining        severe storms

8. In general, barometer readings between 30.00 and 29.80 that are falling mean (circle the correct answer)

           fair weather        it's raining        severe storms

9. In general, barometer readings of 29.80 and below that are falling mean (circle the correct answer)

           fair weather        it's raining        severe storms

10. When the barometer registers 30.10 at 10:00 A.M. and 30.02 at 1:00 P.M., the pressure is (circle the correct answer)

            falling slowly        rising rapidly        falling rapidly

AIR PRESSURE

# STORM PASSAGE IN VAN NUYS, CALIFORNIA (1987)

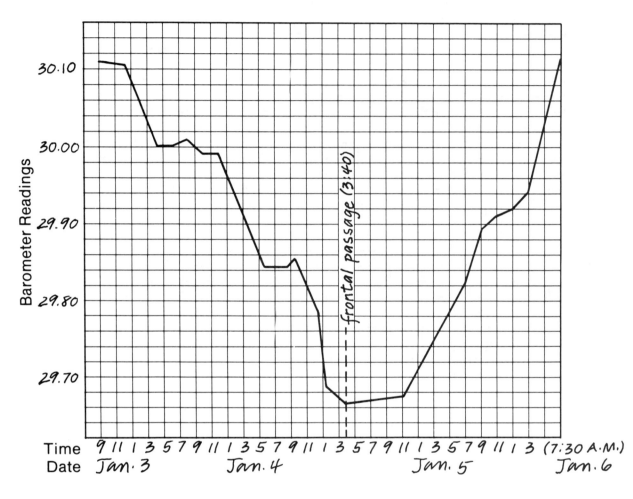

AIR PRESSURE 79

NAME _____

# A STORM'S AIR-PRESSURE PROFILE

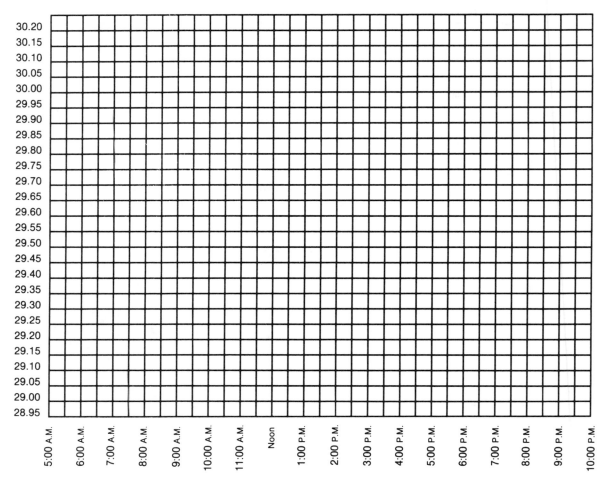

Clouds & Precipitation

Wind Direction

80   AIR PRESSURE

NAME _____

# SUMMER STORM CHART

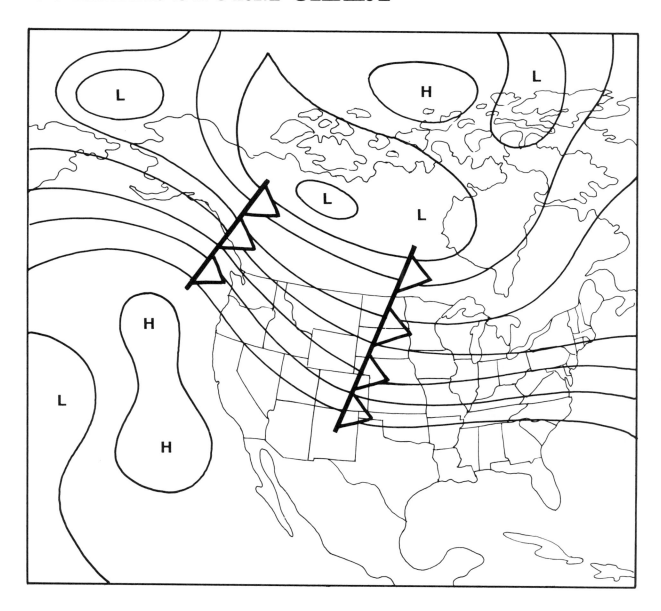

1. Color the high-pressure centers yellow.
2. Color the low-pressure centers blue.
3. Color the storm lines black.
4. What is the weather in your city, as shown on this map?

_____

AIR PRESSURE    81

NAME _____

# Air Pressure Review Test

1. Draw a line between each air pressure level and the kind of weather that is likely to follow.

   29.98 and falling slowly.          Fair

   30.06 and rising slowly.           Possible rain

   29.80 and falling rapidly.         Rain

2. Which of the following readings is average sea-level pressure?

   A. 29.25    B. 29.76    C. 29.92    D. 30.18

3. Give the air pressure reading shown by the arrows on the barometer.

   A. _____ inches    B. _____ inches    C. _____ inches

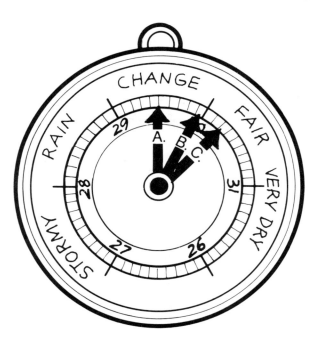

## Air Pressure Review Test continued

4. Label points A, B, and C on the air pressure graph below with one of the following:

- Rain likely
- Frontal passage
- Clearing

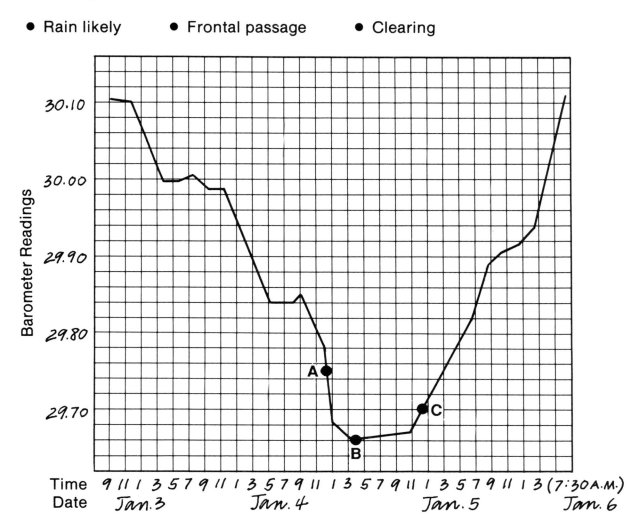

AIR PRESSURE 83

# CHAPTER 8
# FRONTS AND STORMS

A front is a transition between two temperature, or air density, zones. Fronts are formed by a contrast between ocean and land temperatures or by pressure differences in the upper wind flow. These factors cause cold air to sink and warm air to rise. In other words, they cause a front to develop. Frontal weather is unsettled or stormy.

Storms, the most dramatic weather phenomenon, are disturbances of the atmosphere. In addition to frontal storms, there are three major types. Thunderstorms and tornadoes occur in the lower and middle latitudes. Hurricanes (called typhoons over the North Pacific Ocean and willy-willies off the coast of Australia) are tropical storms.

Thunderstorms are the result of massive updrafts of air that create cumulonimbus clouds. The accompanying lightning is caused by the friction of rapidly moving ice particles and rain in the thunderclouds that builds up electrical charges. Thunder is the sound caused by the shock waves of expanding gases moving along the line of electricity. Tornadoes, which are violent whirlpools of air, are caused by the instability of thunderstorms that occur along a cold front. A tornado over water is called a waterspout. Hurricanes are low-pressure cells with extremely high winds. They can develop only when there is a very warm, moist mass of air over the open ocean.

## Concept

The four main types of fronts are the cold front, the warm front, the stationary front, and the occluded front.

## Activity

**Varieties of fronts.** Use page 90, "Types of Weather Fronts," in an opaque projector, or duplicate it and hand it out. Compare the kinds of fronts and the weather they bring.

In a cold front, the cold air cuts under the warm air, forcing the warm air to rise. Cold fronts often produce short periods of heavy precipitation. Colder weather usually follows the passage of a cold front. In the Northern Hemisphere, cold fronts usually lie in a northeast to southwest direction and move toward the east or the southeast. Cold fronts travel about 20 miles an hour, but faster in the winter than in the summer.

In a warm front, warm air overrides the cold air. Warm fronts often produce light, steady precipitation. Warmer weather usually follows the passage of a warm front. In the Northern Hemisphere, warm fronts usually occur on the east side of low-pressure cells and are usually followed by cold fronts as the prevailing westerlies move the low toward the east. Warm fronts travel about 15 miles per hour.

85

# Did you know that . . .

the temperature can suddenly fall dozens of degrees when a front passes! Some historic temperature drops: from 55° at 7:00 A.M. to 8° at 7:15 A.M. in Rapid City, SD (Jan. 10, 1911); from 44° to −56° in 24 hours in Browning, MT (Jan. 23-24, 1916); and from 76° to 10° in 8 hours in Kansas City, MO (Nov. 11, 1911).

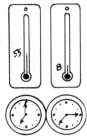

When warm air and cold air meet but move very little, the front is a stationary one. The weather is usually moderate and continues for several days.

The weather of occluded fronts is similar to that of cold or warm fronts but less extreme. Occluded fronts occur because cold fronts travel faster than warm ones. When a cold front catches up with a warm one, it pushes the warm air aloft over the cool air in front of the warm front.

## Concept

A front can be observed locally or located on a weather map by noting sharp temperature changes, drastic changes in humidity, shifts in wind direction, pressure changes, and changes in cloud and precipitation patterns.

## Activities

1. **Weather conditions associated with fronts.** Hand out page 91, "How to Recognize a Front," to your students. Show page 79, "Storm Passage in Van Nuys, California (1987)," on an opaque projector. Is this a warm front or a cold front? What are the signs? What happens as the front passes?

2. **Observing the passage of a front.** Use the worksheet (page 80), "A Storm's Air-Pressure Profile," to chart the passage of a front. (If your students have already done this, use the filled-in worksheet as the basis for class discussion.)

A simple wind shift can indicate a weak front; not all fronts are strong enough to bring dramatic shifts in weather. For a front to be strong, the difference in temperature between the cold front and the warm front must be a large one.

3. **Locating storms on satellite photographs.** Obtain one or more satellite pictures from the nearest National Weather Service office. Show them to your class and discuss the movement of storms that are shown. How are the storms affected by the season (that is, the amount of direct sunlight)? Where are the areas of high and low pressure?

If you have the equipment available, tape the satellite pictures from TV weather reports for a week or two. The tape will be a clear demonstration of the movement of storms.

## Concept

Thunderstorms and tornadoes are storms that occur in the lower middle latitudes. Hurricanes and typhoons are tropical storms.

## Activity

**Storms around the world.** Show the map on page 92 in an opaque projector to illustrate the location of middle-latitude and tropical storms.

Most tornadoes occur in the central Mississippi Valley of the United States, but they can occur in other areas, too.

For a hurricane to survive, the water temperature must be at least 75°-80° and the surface winds must converge. Because hurricanes and typhoons need warm, moist air, they usually begin in late summer or early fall. Cold water off the California coast prevents hurricanes from surviving there. The warm water of the West Atlantic and the Gulf of Mexico creates more favorable conditions for hurricanes.

Notice that the paths of hurricanes in the Northern Hemisphere are west and north, in the Southern Hemisphere, they are east and south. This difference is due to the winds and the rotation of the earth. Hurricanes will diminish over land or over sea when water and air temperatures cool.

During hurricane season, you might trace the paths of hurricanes on a map of the United States as they appear.

## Concept

Lighting and static electricity are caused by the attraction of opposite charges of electricity, which gives off energy.

## Activities

1. **Static electricity (I).** Turn off the room lights and make the room as dark as possible. Unroll cloth friction tape. What do you see?

2. **Static electricity (II).** Turn off the room lights and make the room as dark as possible. Rub two balloons together. What do you see?

3. **Static electricity (III).** Comb your hair (with the lights out for the full effect) and then put the comb in a bowl of puffed rice (and turn the lights on). What do you see? The puffed rice will jump to the "charged" comb, become charged, and then jump off; the cycle will repeat itself.

## Concept

There is a typical pattern of electrical charges in a thundercloud when lightning occurs. Negative charges are concentrated in the middle section of a cumulonimbus cloud. At the bottom are positive zones surrounded by negative zones.

## Activity

**Electrical charges in a thundercloud.** Use the illustration on page 93, "Lightning in a Storm" to show the arrangement of electrical charges in a thundercloud. Opposite electrical charges attract. When the charges build up enough, lightning will occur within a cloud or between the cloud and the ground.

# *Did you know that . . .*

in a severe thunderstorm, more heat is released than is released by several small atomic bombs. And worldwide, there are over 40,000 thunderstorms every day.

## Concept

Objects that conduct electricity are less likely to be damaged by lightning than objects that resist electricity.

## Activities

1. **Lightning rods.** Use the illustration on page 93, "Lightning in a Storm" to show the effect of lightning on various objects and to show the function of a lightning rod. Lightning tends to hit the highest places. If the object is not grounded and has high resistance to electricity (such as a tree or a wooden structure), it may be damaged by the lightning. A lightning rod will attract lightning, and because the rod is grounded by an insulated conducting wire buried in the ground, the lightning will travel down the wire harmlessly into the ground. A grounded TV antenna can serve as a lightning rod.

2. **Lightning safety.** Duplicate the page "Lightning Safety Tips" and hand it out. (If you live in an area of frequent thunderstorms, students may want to take this sheet home to share with their family.) Discuss students' experiences with lightning and thunderstorms. Discuss the reasons for the safety measures: for example, lightning will hit the highest spot; the electrical charge will pass through materials that are conductive, such as metal and water; the electrical charge can damage materials that are resistant, such as wood.

## Concept

Hurricane winds, like winds of other low-pressure systems, move in a counter-clockwise direction north of the equator.

## Activity

**How a hurricane travels.** Use page 95, "A Bird's-Eye View of a Hurricane," to illustrate the structure and movement of a hurri-cane. The wind pushes the hurricane west and north (when it is north of the equator); if the hurricane is strong enough, it may finish by moving east (especially across the Atlantic Ocean). The winds swirl around the "eye," where the air is calm.

## Concept

The lower the air pressure and the faster the wind, the more damage a hurricane will cause.

## Activity

**Rating the damage potential of a hurricane.** Page 96 is a simplified version of the Saffir-Simpson Hurricane Damage-Potential Scale. During hurricane season, monitor the news descriptions of hurricanes. Have your students use this scale to predict the damage potential of a hurricane from the wind speeds reported in the newspaper.

Hurricane eyes travel at 10 to 30 miles per hour. Winds in the wall cloud area can blow at 130 to 150 miles per hour. It is these winds that cause the destruction.

## Concept

Tornadoes develop along squall lines of cumulonimbus clouds, most frequently in the Midwest and the states that border the Gulf of Mexico.

## Activities

1. **Tornado incidence by state.** Tornadoes occur most frequently in the central United States, especially the lower Mississippi River Valley. Show the map on page 98 on an opaque projector. Which state has the most tornadoes? (The numbers are an annual average.) What is the average number of tornadoes in your state?

88    THE WEATHER REPORT

**2. Features of a tornado.** Show the page "A Terrible Twister" on an opaque projector to illustrate the features of a tornado. The winds of a tornado are the most violent winds on earth; they may whirl around at speeds of 100–200 miles per hour and even reach 300 miles per hour on occasion. Because no instrument can be placed in a tornado, however, these speeds cannot be measured; they can only be estimated. Scientists estimate the wind speed of a tornado according to its size, movement, and the damage it causes. They make these estimates by direct observation or by studying films of the tornado.

Dark pouches on the underside of cumulonimbus clouds, called mammatus clouds, often signal a tornado. The tornado is a funnel cloud rotating counterclockwise (in the Northern Hemisphere) that comes out of the bottom of heavy, dark clouds. The tornado moves along the earth at average speeds of 20 to 40 miles per hour. The 100- to 200-mph updraft at at the center of the funnel can suck up houses, animals, and cars. Tornadoes tend to develop in spring or early summer (and occasionally in the fall) on hot, humid days in the afternoon or early evening.

## ANSWER KEY:
## Front and Storms Review Test

**1.** Cold front: Heavy showers. Warm front: Light rain. Occluded Front: Possible showers. **2.** C. **3.** B. and C. **4.** The water temperatures in East Coast waters are warmer. **5.** C. **6.** Instruments cannot be placed into the tornado where the winds are the strongest. **7.** B. **8.** Stay low, away from high exposed ground. Get away from water. Don't touch metal. Make yourself small by dropping to your knees and bending forward and putting your hands on your knees.

# TYPES OF WEATHER FRONTS

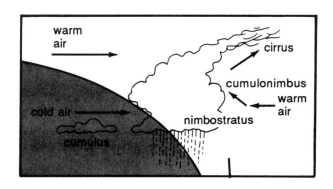

## Cold Front

Cold air mass moves under warm air mass.

Cold front brings brief, heavy precipitation, and then colder weather.

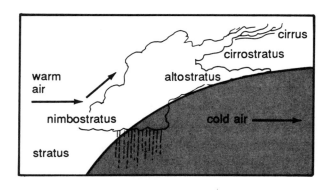

## Warm Front

Warm air mass moves up slope of cold air mass.

Warm front brings steady, light precipitation, and then warmer weather.

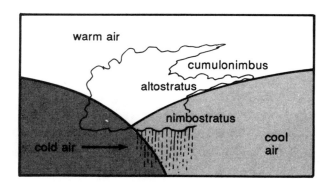

## Occluded Front

Cold air mass meets cool air mass under warm air.

Occluded-front weather is more moderate than cold-front or warm-front weather.

FRONTS AND STORMS

# HOW TO RECOGNIZE A FRONT

## Cold Front (sharp contrast or change in weather)

| Weather Element | Before Passing | While Passing | After Passing |
|---|---|---|---|
| Winds | south-southwest | gusty, shifting | west-northwest |
| Temperature | warm or mild | sudden drop | colder |
| Pressure | falling steadily | sharp rise | rising steadily |
| Clouds | increasing cirrus and cirrostratus, then either towering cumulus or cumulonimbus | towering cumulus or cumulonimbus | often cumulus |
| Precipitation | short period of showers | heavy rain or snow showers, sometimes with hail, thunder, and lightning | decreasing intensity of showers, then clearing |
| Humidity | high; remains steady | sharp drop | lowering |

## Warm Front (gradual change in weather)

| Weather Element | Before Passing | While Passing | After Passing |
|---|---|---|---|
| Winds | south-southeast | variable | south-southwest |
| Temperature | cool or cold | steady rise | warmer |
| Pressure | falling | leveling off | slight rise, then fall |
| Clouds | cirrus, cirrostratus, altostratus, nimbostratus, stratus; fog and cumulonimbus in summer | stratus-type | clearing with scattered stratocumulus; cumulonimbus in summer |
| Precipitation | light to moderate rain, snow, sleet, or drizzle | drizzle | usually none; sometimes, light showers |
| Humidity | (change in humidity for warm front is insignificant) | | |

*The Weather Report* © 1989

FRONTS AND STORMS 91

# TEMPERATE AND TROPICAL STORMS

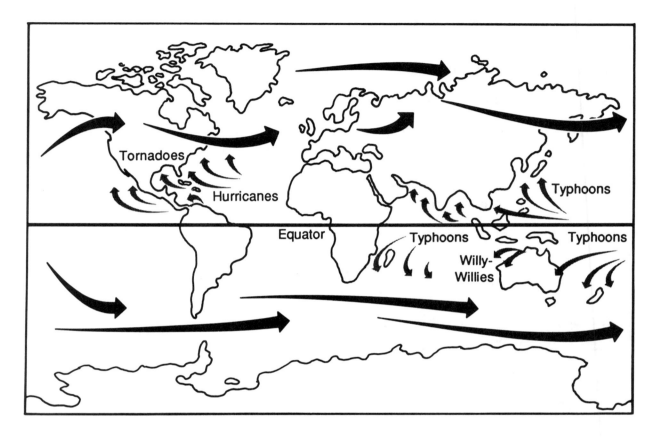

Paths of Tropical Storms

Paths of Middle Latitude Storms

92    FRONTS AND STORMS

# LIGHTNING IN A STORM

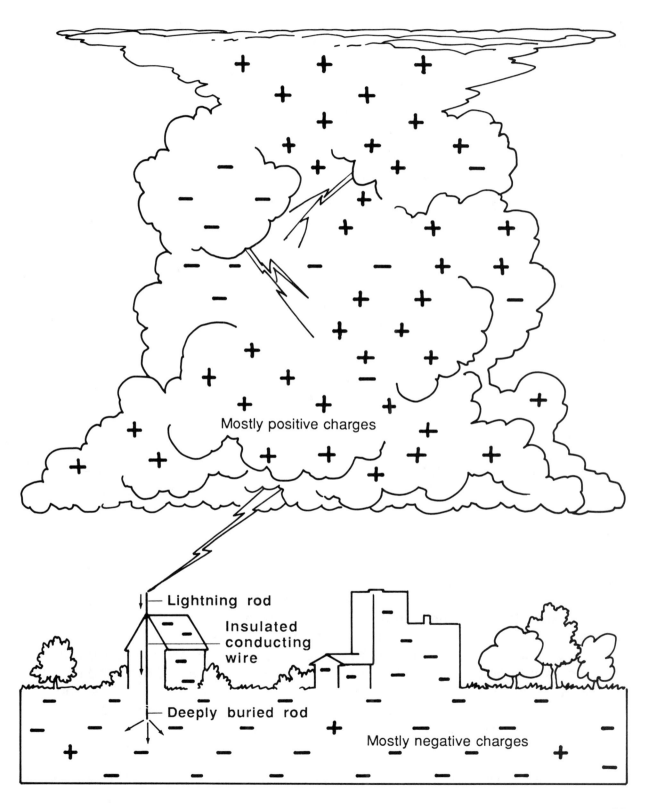

FRONTS AND STORMS

# LIGHTNING SAFETY TIPS

When a thunderstorm is approaching, stay inside or seek shelter in a home, a large building, or a car (all-metal, not a convertible). Use the telephone only for an emergency.

If you are caught outside in a thunderstorm, follow these rules:

1. Stay low. Don't stand on a hilltop; go to a valley or a ravine or under a thick growth of small trees.

2. Get off or away from open water. Don't plan a day of sailing or swimming if a thunderstorm has been predicted.

3. Don't touch metal. Get off motorcycles, bicycles, golf carts, and tractors. Put down golf clubs and take off golf shoes or other sports shoes with metal cleats. Stay away from wire fences, clotheslines, metal pipes, and rails.

4. Make yourself small. If you are in a group, spread out. If you are out in a level field or prairie and feel your hair stand on end, lightning may be about to strike you so make yourself as small a target as possible. Drop to your knees and bend forward, putting your hands on your knees. Do not lie flat on the ground.

## First Aid

People who have been struck by lightning carry no electrical charge and can be handled safely. They have received a severe electrical shock and may be burned. If a lightning victim is not breathing, immediately start mouth-to-mouth resuscitation. If the victim is not breathing and has no pulse, cardio-pulmonary resuscitation (CPR) should be administered as soon as possible by someone who has been trained in the technique. Victims who appear only stunned should be checked for burns, especially at fingers and toes and next to buckles and jewelry. Give first aid for shock and send for help; do not let the victim walk around. The American Red Cross has more information about how to help a person who has been struck by lightning.

---

*Source:*
"Thunderstorms and Lightning," National Oceanic and Atmospheric Administration, NOAA/PA 83001, revised June 1985.

# A BIRD'S-EYE VIEW OF A HURRICANE

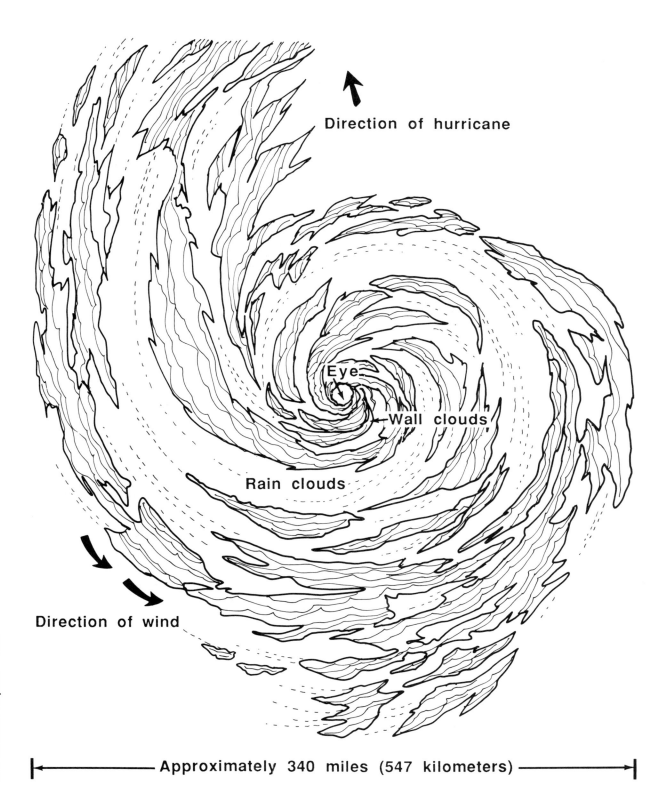

# HURRICANE DAMAGE-POTENTIAL SCALE

| Scale Number | Central Pressure (inches) | Winds (mph) | Damage |
|---|---|---|---|
| 1 | 28.94 and above | 74–95 | damage mainly to trees, shrubbery, and unanchored mobile homes |
| 2 | 28.50–28.91 | 96–110 | some trees blown down; major damage to exposed mobile homes; some damage to roofs of buildings |
| 3 | 27.91–28.47 | 111–130 | foliage removed from trees; large trees blown down; mobile homes destroyed; some structural damage to small buildings |
| 4 | 27.17–27.88 | 131–155 | all signs blown down; extensive damage to roofs, windows, and doors; complete destruction of mobile homes; flooding inland as far as 6 miles; major flood damage to lower floors of structures near shore |
| 5 | below 27.17 | above 155 | severe damage to windows and doors, extensive damage to roofs of homes and industrial buildings; small buildings overturned and blown away; major flood damage to lower floors of all structures less than 15 feet above sea level within 500 miles of shore |

*The Weather Report* © 1989

# A TERRIBLE TWISTER

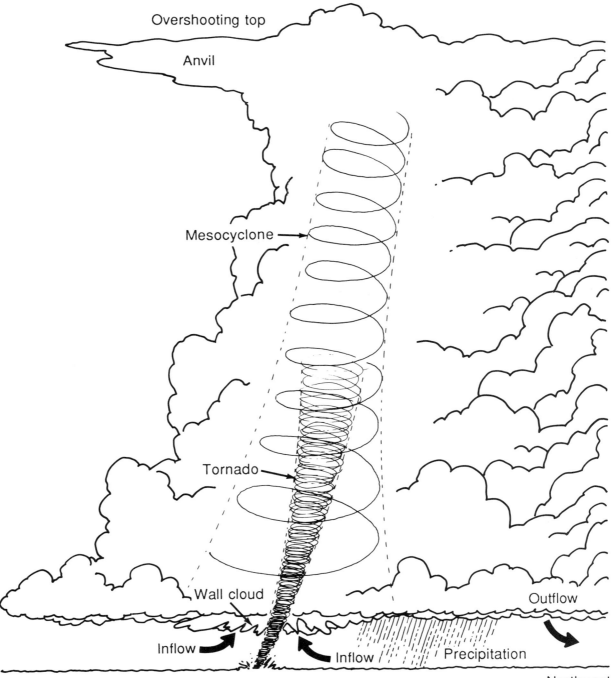

Some of the features associated with a tornado-breeding thunderstorm as viewed from the southeast. The thunderstorm is moving to the northeast. The tornado forms in the southwest part of the thunderstorm. A mesocyclone is a small but violently spinning column of air within a cumulonimbus cloud.

FRONTS AND STORMS

# TORNADO COUNTRY

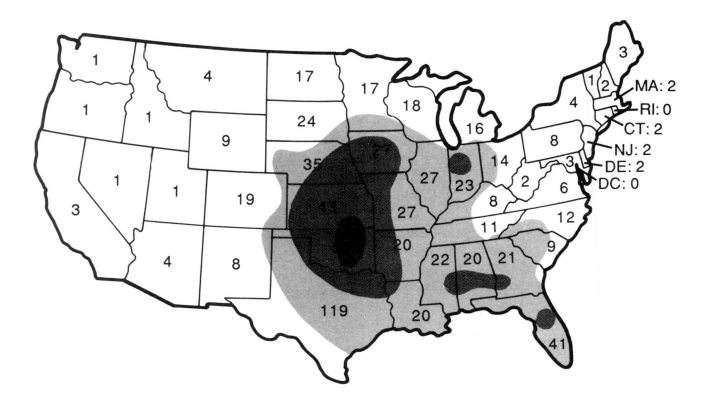

The darker the shading, the greater the frequency of tornadoes. The number in each state is the average number of tornadoes each year between 1953 and 1980.

NAME _____

# Fronts and Storms Review Test

1. Draw a line between each type of front and the weather that accompanies it.

   Cold front                        Possible showers

   Warm front                        Heavy showers

   Occluded front                    Light rain

2. Which of the following temperature changes would most likely occur during passage of a cold front in Missoula, Montana?

   A. 58° at 11 A.M.; 57° at 2 P.M.     B. 58° at 11 A.M.; 64° at 2 P.M.
   C. 58° at 11 A.M.; 45° at 2 P.M.

3. Which of the following would not indicate the passage of a cold front?

   A. Rising air pressure        C. Falling air pressure
   B. Decreasing showers         D. A shift in wind direction

4. Why do hurricanes occur more often off the East Coast than off the West

   Coast of the United States? _____

   _____

5. In which one of the following states do tornadoes occur most often?

   A. California     B. New York     C. Oklahoma     D. North Dakota

6. Tornado winds have been clocked up to 200 miles per hour, but many may have even stronger winds. Why haven't they been clocked at higher speeds?

   _____

7. When are tornadoes most likely to occur?

   A. A winter morning       C. A winter evening
   B. A summer evening       D. A summer morning

8. What are four precautions a person should take when a thunderstorm is approaching?

   A. _____

   B. _____

   C. _____

   D. _____

*The Weather Report* © 1989

FRONTS AND STORMS     99

# CHAPTER 9
# TOPOGRAPHY AND WEATHER

Air temperature decreases with altitude. The reasons are that (1) the collisions between air molecules decrease at higher elevations, and (2) air heated by the earth's surface rises and mixes, moving the heat energy higher, so the farther a spot is above the earth's surface, the less heat will reach it.

Precipitation is greater on the windward than on the leeward side of mountains. This phenomenon is called the rain shadow effect. As air rises, its pressure drops, causing it to cool. As air descends, its pressure increases, causing it to warm. The cooling causes water vapor to condense into clouds and precipitation.

## Concept

Air temperature becomes cooler the higher you go; it drops an average of 3.6° for every 1,000 feet of elevation above sea level.

## Activities

1. **Comparing readings.** Place a thermometer on the ground outside, another one 5 feet high, and a third one 10 feet high, if possible. Check the temperatures 10 minutes later and compare them.

2. **Heat from the earth's surface.** Discuss the effects of radiational heating of the earth's surface. The sunlight heats the ground, and the air near the ground receives that heat through conduction. Air is a poor conductor of heat, however, so only the air nearest the ground receives much of this heat. As the day progresses, rising air bubbles (called thermals) help transfer some of this heat to higher levels.

   At night the ground and the air near it cool off by radiating infrared energy. Because this radiation cooling occurs at ground level, there is less temperature variation at higher elevations. In fact, at night there is a temperature inversion at an elevation of about 100 to 200 feet; the air here is much warmer than the air at ground level because it is not affected by surface cooling. Because of this temperature inversion, helicopters are used to prevent frost damage to crops; by circling above the fields, they can mix the warmer air with the cool nighttime air near the ground.

   Hand out the worksheet "Hot and Cool in a High-Rise." After your students finish it, discuss the answers.

3. **High altitudes and "thinner" air.** Ask your students if any of them have ever gone mountain climbing. How does being at a higher elevation affect your breathing? Note that there is less oxygen in the air at higher elevations. How does the temperature change as you climb?

   Hand out the worksheet "High Spots and Low Spots." Discuss the answers to questions 2, 3, and 4 in class.

# Did you know that...

on a clear, windless day, if the temperature at a jogger's feet is 50°, the temperature at waist level may be only 32°.

## ANSWER KEY:
### Hot and Cool in a High-Rise

1. 7.5° (between 62.5° and 70°). **2.** 2° (between 55° and 57°). **3.** The maximum temperature varies more. Because sunlight heats the ground, and the ground is a better conductor of heat than the air. **4.** 0 (ground level): 15°; 330 feet: 10°; 650 feet: 9°; 980 feet: 7.5°. The greatest temperature range is at ground level. The ground absorbs more heat from the sun during the day than air does, and it loses more heat at night by radiating infrared energy.

## ANSWER KEY:
### High Spots and Low Spots

1. Singapore: 81; Kansas City: 64; Quito: 55.
2. Quito. It is the highest in elevation.
3. Singapore. It is the lowest in elevation.
4.

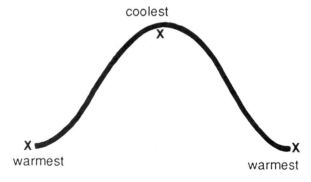

## Concept

Areas on the windward side of mountains have more precipitation than areas on the leeward side of mountains. This is called the rain-shadow effect.

## Activities

1. **Mountains and rainfall.** Show the page "Mountains and Rainfall" in an opaque projector to explain the rain-shadow effect. Mountains affect precipitation because clouds strengthen on the windward side (in most of North America, this is the western side) of mountains. When a current of warm, moist air reaches a mountain, it is forced to rise. As it rises, the pressure of the air decreases, causing the air to cool. This cooling causes water vapor to condense into clouds and precipitation. Thus, when the air travels down the leeward side of a mountain (in most of North America, the eastern side), it has lost much of its moisture.

2. **The dry side and the wet side.** To show the effect of mountains on precipitation, give your students the worksheet "Seattle vs. Yakima." These two cities provide an excellent example of how mountains affect rainfall. Seattle is on the western side of the Cascades and Yakima is on the eastern side. Discuss the worksheet answers with the whole class.
   
   To help your students see the contrast more clearly, have them complete the graph on the worksheet "Rainfall Graph for Seattle and Yakima."

3. **Rainy peaks and dry valleys.** As another demonstration of the effect of mountains on precipitation, hand out the worksheet "Up Mountains and Down into Valleys." Your students should first mark a vertical line for the rainfall level for each location, as has been done for

102    THE WEATHER REPORT

the first three rainfall levels. When they have marked each rainfall level, they should connect the top of each vertical line (the rainfall-level points) with a dotted horizontal line. How does the pattern of rainfall compare to the pattern of mountains and valleys?

4. **Imaginary cities.** To reinforce learning about the rain-shadow effect, have your students draw an imaginary state with mountain ranges and cities (have them name the mountains and cities, too, for fun). Then they should show where the rainfall will be the lightest and the heaviest according to the rain-shadow effect.

## ANSWER KEY:
### Seattle vs. Yakima

1. Seattle: 38.60"; Yakima: 7.98".
2. Seattle is on the western side (the windward side) of the Cascade Mountains, and Yakima is on the eastern side (the leeward side) of the mountains. 3. Seattle: December; Yakima: January. 4. Seattle: July; Yakima: July. 5. Answers will vary.

## ANSWER KEY:
### Rainfall Graph for Seattle and Yakima

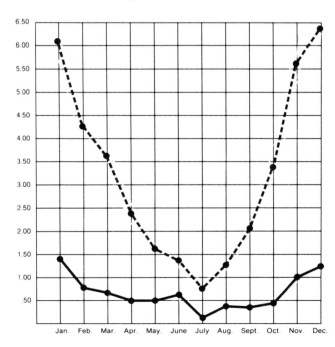

## ANSWER KEY:
### Up Mountains and Down into Valleys

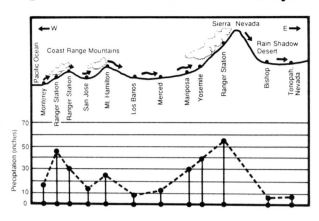

## ANSWER KEY:
### Topography and Weather Review Test

1. True.  2. False.

3.

4.

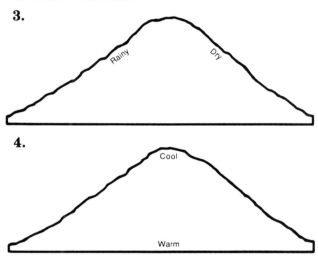

**TOPOGRAPHY AND WEATHER**    103

NAME _____

# HOT AND COOL IN A HIGH-RISE

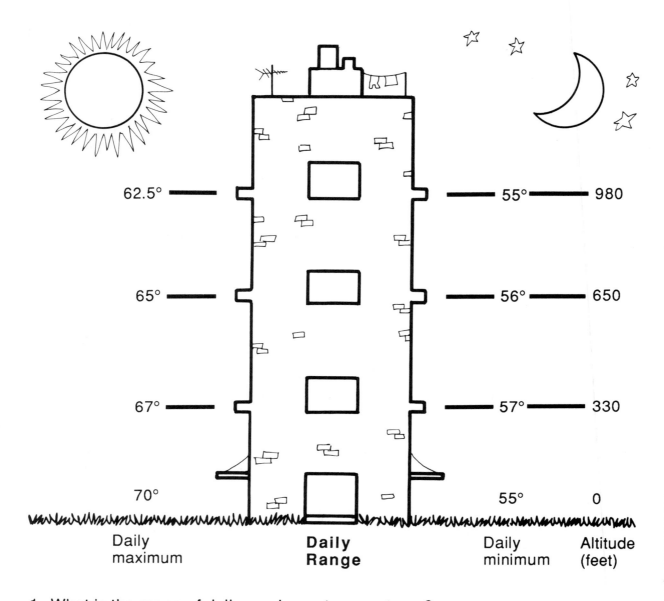

1. What is the range of daily maximum temperatures? _____
2. What is the range of daily minimum temperatures? _____
3. Does the maximum or the minimum temperature vary more? _____
   Why do you think this is so? _____
4. Put the range for each altitude in the door or window square. At what altitude is there the greatest temperature range? _____
   Why do you think this is so? _____

104  TOPOGRAPHY AND WEATHER

NAME _____

# HIGH SPOTS AND LOW SPOTS

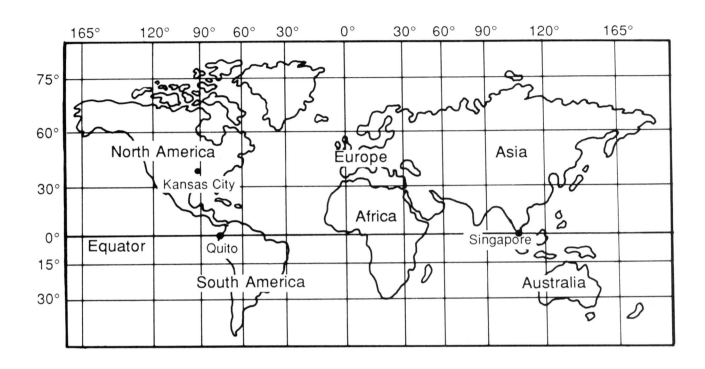

1. Find the yearly average temperatures for the three cities and fill them in on the chart.

2. Which city is coldest overall?

   _____

   Why do you think this is so?

   _____

3. Which city is warmest overall?

   _____

   Why do you think this is so?

   _____

4. On a separate sheet of paper, draw a mountain. Label the coolest and warmest parts.

**Average Monthly Mean Temperature**

|  | Singapore, Malaya (10 feet) | Kansas City, Missouri (750 feet) | Quito, Ecuador (9350 feet) |
|---|---|---|---|
| January | 80 | 35 | 55 |
| February | 80 | 41 | 55 |
| March | 81 | 51 | 55 |
| April | 82 | 65 | 55 |
| May | 82 | 75 | 55 |
| June | 81 | 83 | 55 |
| July | 81 | 89 | 55 |
| August | 81 | 87 | 55 |
| September | 81 | 79 | 55 |
| October | 81 | 68 | 55 |
| November | 81 | 52 | 54 |
| December | 81 | 40 | 54 |
| Yearly Average |  |  |  |

TOPOGRAPHY AND WEATHER

# MOUNTAINS AND RAINFALL

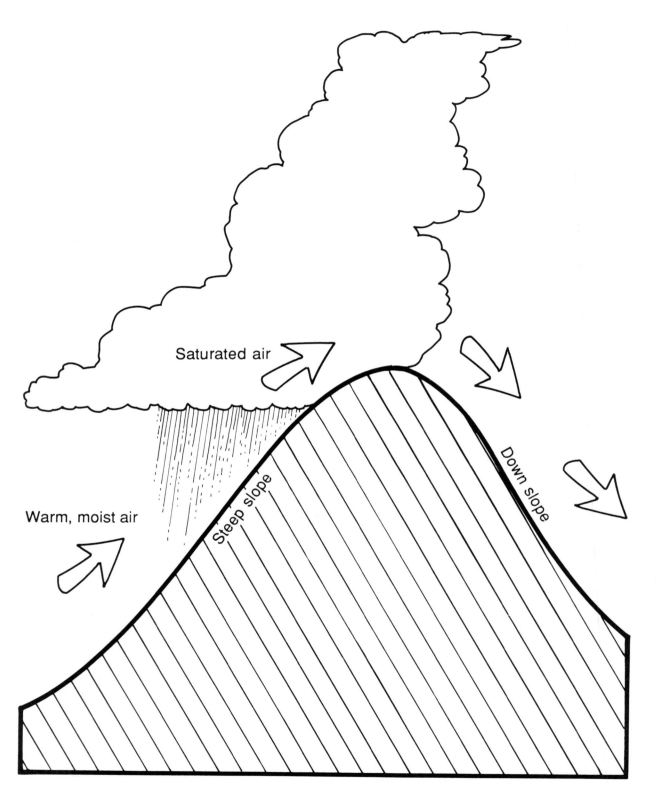

106  TOPOGRAPHY AND WEATHER

NAME _____

# SEATTLE VS. YAKIMA

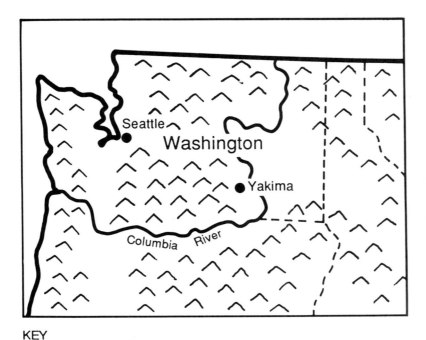

| | Average Inches of Rainfall | |
|---|---|---|
| | Seattle | Yakima |
| January | 6.04 | 1.44 |
| February | 4.22 | 0.74 |
| March | 3.59 | 0.65 |
| April | 2.40 | 0.50 |
| May | 1.58 | 0.48 |
| June | 1.38 | 0.60 |
| July | 0.74 | 0.14 |
| August | 1.27 | 0.36 |
| September | 2.02 | 0.33 |
| October | 3.43 | 0.47 |
| November | 5.60 | 0.97 |
| December | 6.33 | 1.30 |
| Yearly Average | | |

KEY

- - - state border          ▢ plains

⋀ ⋀ mountains          ~ river

1. Find the yearly average rainfall total for Seattle and Yakima and add them to the chart.

2. Why does Seattle get more rain than Yakima? _____

3. What is the wettest month of the year for Seattle? _____

   What is the wettest month of the year for Yakima? _____

4. What is the driest month of the year for Seattle? _____

   What is the driest month of the year for Yakima? _____

5. Name some wet and some dry areas in your state: _____

   _____

   Do any fit the same pattern as Seattle and Yakima? _____

   Which ones? _____

TOPOGRAPHY AND WEATHER

NAME _____

# RAINFALL GRAPH FOR SEATTLE AND YAKIMA

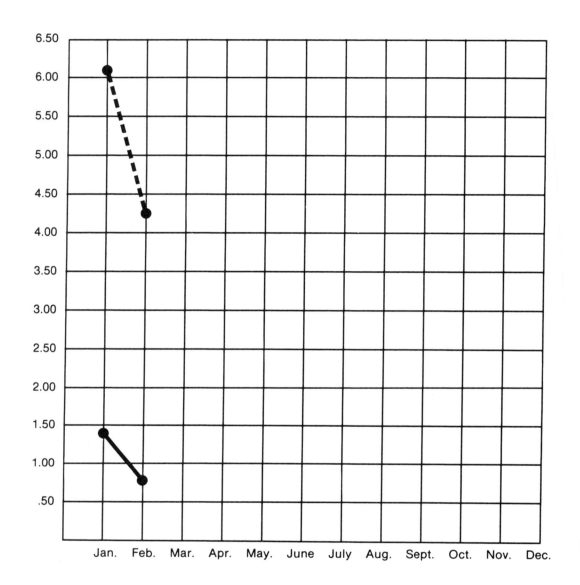

Use the numbers of the "Seattle vs. Yakima" worksheet to fill in this graph.

1. Mark the points for each month for Seattle.
2. Connect Seattle's points with a dotted line.
3. Mark the points for each month for Yakima.
4. Connect Yakima's points with a solid line.

January and February have already been marked for you.

TOPOGRAPHY AND WEATHER

NAME _____

# UP MOUNTAINS AND DOWN INTO VALLEYS

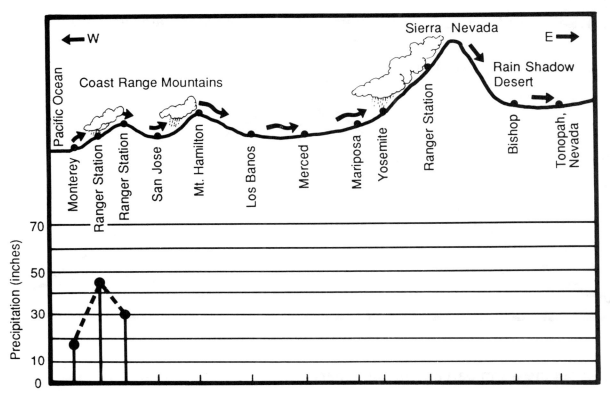

Graph the yearly precipitation averages for this cross section of California, using the rainfall averages below. The first three points have already been done for you.

## Annual Precipitation Averages

San Jose: 13 inches

Mt. Hamilton: 25 inches

Los Banos: 9 inches

Merced: 12 inches

Mariposa: 30 inches

Yosemite: 38 inches

Ranger station: 55 inches

Bishop: 6 inches

Tonopah, Nevada: 8 inches

TOPOGRAPHY AND WEATHER

NAME _____

# Topography and Weather Review Test

1. True \_\_\_ or False \_\_\_ . The higher you go above 200 feet, the cooler it gets.

2. True \_\_\_ or False \_\_\_ . The eastern slopes of most mountains in the United States receive more rain.

3. Label the rainy part and the dry part of this mountain.

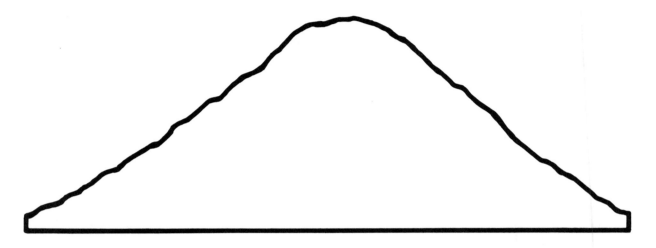

4. Label the warm part and the cool part of this mountain.

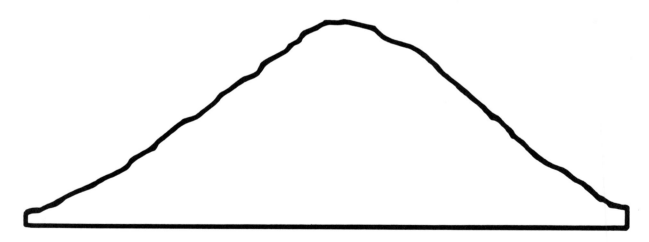

110    TOPOGRAPHY AND WEATHER

# CHAPTER 10
# GEOGRAPHY AND CLIMATE

Latitude, or distance from the equator, affects the range and seasonal variation of temperature. The closer a spot is to the equator, the more direct sunlight it gets, so it is warmer and has less seasonal variation in temperature than a spot farther from the equator.

Proximity to the ocean also affects the temperature of coastal areas. They are warmer in winter and cooler in summer than inland areas equally far from the equator, because water takes longer to heat than land does. The effect of water temperature on land temperature is more pronounced if the ocean (or other large body of water) is to the west, because the prevailing wind over most of the earth's land masses moves from west to east.

The air is more humid at the equator because there is more evaporation than at other latitudes. Precipitation and the saturation point of the air depend on topography (see the chapter on Topography and Weather), temperature, and the direction of the prevailing winds and the movement of pressure systems.

In summary, the climate of any given spot is the result of its topography and its geographical location interacting with the sun, wind, and moisture in the air.

## Concept

The closer a place is to the equator, the more direct sunlight it gets, so it is warmer and has less seasonal variation in temperature than a place farther from the equator.

## Activities

1. **Latitude and sunlight (I).** In a dark room, shine a flashlight on the equator of a globe. Discuss the differences in the brightness of the light at the equator and toward the poles. The more direct the sunlight, the warmer the temperature.

   To show the seasonal change of sunlight, shine the flashlight a little above the equator to represent the position of the sun when it is summer in the Northern Hemisphere and winter in the Southern Hemisphere. Shine the flashlight a little below the equator to show the position of the sun when it is winter north of the equator and summer to the south. Discuss how temperatures might vary at different locations.

2. **Latitude and sunlight (II).** Show page 116, "Latitude and Sunlight," in an opaque projector to reinforce and further explain the demonstration with the globe in 1, above.

3. **Direct sun = higher temperatures.** Have your students work in pairs for this experiment. One will hold a flashlight 2 centimeters above a piece of graph paper while the other draws the outline of the light reflected on the

paper. First, the flashlight should be held perpendicular to the surface of the paper, and then it should be held at an angle. Have the students count the number of squares in each circle. Which way covers more squares? Which way is the light dimmer on the paper? The perpendicular beam is brighter and covers fewer squares; it represents direct sunlight, which brings warm temperatures. The beam of light at an angle covers more squares but the light is dimmer; it is the same amount of light as the perpendicular beam but it is less concentrated. The beam at an angle represents indirect sunlight, which brings cooler temperatures. Direct sunlight occurs at the equator; sunlight becomes more indirect the nearer to the poles it is.

4. **Latitude and average annual temperatures.** Use an opaque projector to show the graph "Average Annual Sea-Level Temperatures by Latitude" (page 119). Ask your students to find the approximate annual temperature where they live.

5. **Latitude, temperature level, and seasonal variation.** Hand out the activity sheet "Hot Spots and Cold Spells" (page 117) to help your students review and use the concept that the farther a place is from the equator, the cooler it will be overall (but not always every day) and the more the temperature will vary seasonally.

ANSWER KEY:
## Hot Spots and Cold Spells

1. Moscow. **2.** Singapore. **3.** Singapore. Because it is closest to the equator.
4. January. **5.** July. **6.** The temperature averages are lower and the seasonal variations are greater.

## Concept

Because water heats and cools more slowly than land, the temperature range of coastal land areas is smaller than that of inland areas. This effect is particularly noticeable if the prevailing winds blow from the body of water toward the land.

## Activities

1. **Soil heats and cools more quickly than water.** Fill one coffee can with water, and fill another one with soil. Place both in a refrigerator until they are the same temperature. As they are chilling, test them occasionally. Which can cools more quickly?

   When both cans are the same temperature, place them in the sun. Which one heats more quickly? How does the difference in the capacity to absorb and release heat affect the temperature of land near large bodies of water?

2. **Water temperature affects land temperature.** Hand out the worksheet "Seaside or Landlocked?" (page 120). The average monthly temperatures are the average of the high and low temperatures for the month. Discuss the answers as a class. Is the temperature of your city affected by proximity to an ocean or other large body of water?

ANSWER KEY:
**Seaside or Landlocked?**

**1.–3.** See map.

|  | Average Monthly Mean Temperature (°F) |  |  |  | Average Temp. Range |
|---|---|---|---|---|---|
|  | January | April | July | October |  |
| Los Angeles, CA (34° N) | 56 | 60 | 69 | 66 | *13* |
| Phoenix, AZ (33°30' N) | 52 | 68 | 92 | 73 | *40* |
| Dallas-Ft. Worth (33° N) | 44 | 66 | 86 | 68 | *42* |
| Birmingham, AL (33°30' N) | 43 | 63 | 80 | 63 | *37* |
| Charleston, SC (33° N) | 48 | 64 | 80 | 66 | *32* |

**4.** Los Angeles. Because it is on the coast and the prevailing wind blows from the ocean to the land. **5.** Charleston. Because it is on the coast, but the winds blow from the land toward the ocean. **6.** The temperatures of the inland city are hotter in the summer and cooler in the winter. Because water heats and cools more slowly than land, the wind that blows from the ocean will cool the land in summer and warm it in winter.

GEOGRAPHY AND CLIMATE

## Did you know that...

the air of the deserts of California and Arizona is often more humid than the air at the North Pole? This is because desert air is warmer than polar air and therefore can hold more water before it is saturated. Antarctica is often referred to as a polar desert.

## Concept

Average temperature is affected by latitude and by proximity to an ocean.

## Activity

**Temperature highs and lows.** Duplicate the table and the questions about "Temperature Highs and Lows Around the World." Use the questions as the basis for class discussion or have students answer them by themselves, with the help of a world map.

ANSWER KEY:
**Temperature Highs and Lows Around the World: Questions**

1. Libya.  2. 100°; -80°; 180°.  3. 100°; 14°; 86°.  4. Alaska; Hawaii is closer to the equator than Alaska and therefore has less seasonal temperature variation; Hawaii is surrounded by water, which helps to keep its overall temperature warmer than Alaska's.

## Concept

Because warm air holds more water vapor than cold air (see page 24), the air at the equator is more humid than the air at the poles.

## Activity

**Humidity at the equator and at the poles.** Use page 123, "Average Humidities by Latitude" in an opaque projector to explain this concept to your students. Ask them to locate their approximate latitude on the graph and to estimate the average specific humidity (grams of water vapor per kilogram of air).

## Concept

The amount of precipitation in an area is influenced by topography, direction of prevailing winds, movement of air pressure systems, average temperatures, and proximity to large bodies of water.

## Activities

1. **Wet and dry in your state.** To help your students understand influences on precipitation patterns in your state, obtain a topographical map of your state and a record of average precipitation for various locations in your state (see Appendix or consult the nearest National Weather Service office). Locate the spot with the highest average precipitation and the spot with the lowest. Are the precipitation levels in these two spots affected by terrain? By temperature? By proximity to large bodies of water? Is the precipitation level in your city affected by any of these factors.

114   THE WEATHER REPORT

**2. Global precipitation patterns.** Duplicate the table and the questions "Precipitation Highs and Lows Around the World" and hand them out to your students. Use the questions to discuss influences on precipitation. (Remember that the windward side of mountains varies according to the prevailing wind at various latitudes; see page 66, "Global Wind Patterns.") Refer to the graph "Average Humidities by Latitude (page 123).

## ANSWER KEY:
### Precipitation Highs and Lows Around the World: Questions

**1.** They are surrounded by the ocean; they are in latitudes of fairly high humidity. **2.** In the western mountains—the Cascades, the Sierras, and the Rockies. *(Any one of the following)* The prevailing westerly winds carry moisture from the Pacific Ocean; the humidity from the ocean is pushed up these high western mountain ranges, where it condenses as precipitation. **3.** *(Any two of the following)* There are no mountains to cause humid air to rise and condense as precipitation; the prevailing winds are southwest, blowing from land to ocean, so they do not carry much humidity; the prevailing winds are the trade winds, which are gentle and do not bring storms.

**4.** Cherrapunji is in monsoon country. In the summer the land becomes much warmer than the ocean, creating lower air pressure inland and causing the wind to shift direction and blow inland from the sea. This suddenly brings heavy rains and high humidity. Monsoons move into India from the southwest and "retreat" at the end of the season in the same direction, so the southwest, where Cherrapunji is located, gets more monsoon rain than any other part of India. In addition, Cherrapunji is in the foothills on the windward side of the Himalaya Mountains.

## ANSWER KEY:
### Geography and Climate Review Test

**1.** B and D. **2.** C. **3.** Warmest: Nairobi. Coolest: New York City. **4.** San Francisco.

# LATITUDE AND SUNLIGHT

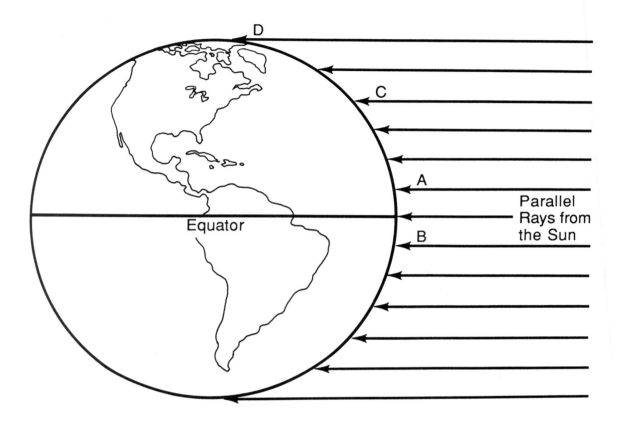

Because the earth is curved, sunlight hits the equator directly but it hits the poles at an angle. The same amount of sunlight that hits an area at the equator (A to B) covers a much larger area near the North Pole (C to D).

NAME _____

# HOT SPOTS AND COLD SPELLS

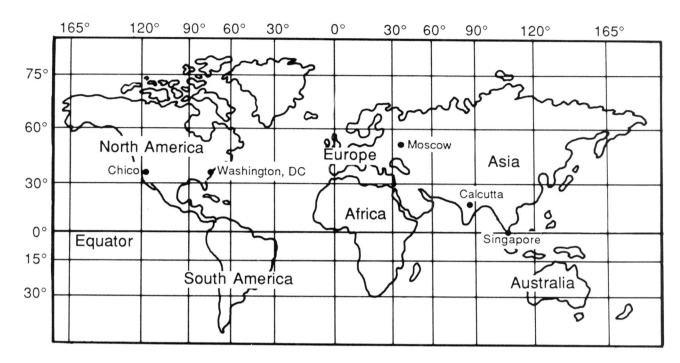

**Average Monthly Mean Temperature (°F)**

|  | Singapore, Malaya (1° North latitude) | Calcutta, India (23° North latitude) | Washington, D.C. (39° North latitude) | Moscow, USSR (56° North latitude) | Chico, Calif. (39° North latitude) |
| --- | --- | --- | --- | --- | --- |
| January | 80 | 67 | 36 | 12 | 55 |
| February | 80 | 71 | 37 | 15 | 60 |
| March | 81 | 80 | 46 | 24 | 60 |
| April | 82 | 85 | 55 | 38 | 71 |
| May | 82 | 86 | 65 | 53 | 81 |
| June | 81 | 85 | 74 | 62 | 89 |
| July | 81 | 84 | 78 | 66 | 95 |
| August | 81 | 83 | 76 | 63 | 94 |
| September | 81 | 83 | 70 | 52 | 90 |
| October | 81 | 81 | 58 | 40 | 78 |
| November | 80 | 73 | 48 | 28 | 64 |
| December | 80 | 67 | 38 | 17 | 57 |
| Yearly Average | 81 | 79 | 57 | 39 | 74 |

GEOGRAPHY AND CLIMATE

# Hot Spots and Cold Spells continued _____

1. Which city is coldest overall? _____

2. Which city is warmest overall? _____

3. In which city does the temperature stay nearly the same all year?

   _____ Why? _____

4. What is the coldest month for all five cities? _____

5. What is the warmest month in Moscow? _____

6. How do temperature averages and seasonal variations change as you move

   farther from the equator? _____

   _____

   _____

*The Weather Report © 1989*

# AVERAGE ANNUAL SEA-LEVEL TEMPERATURES BY LATITUDE

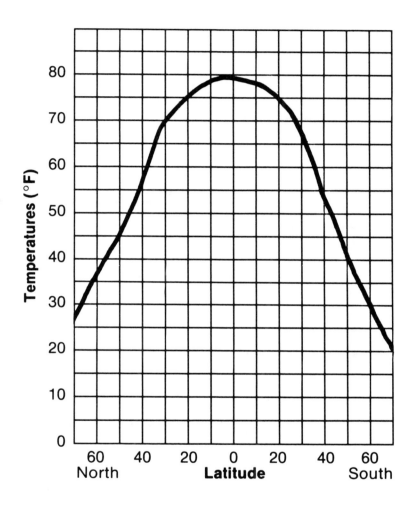

GEOGRAPHY AND CLIMATE 119

NAME _____

# SEASIDE OR LANDLOCKED?

|  | Average Monthly Mean Temperature (°F) ||||  Average Temp. Range |
|---|---|---|---|---|---|
|  | January | April | July | October |  |
| Los Angeles, CA (34° N) | 56 | 60 | 69 | 66 |  |
| Phoenix, AZ (33°30′ N) | 52 | 68 | 92 | 73 |  |
| Dallas–Ft. Worth (33° N) | 44 | 66 | 86 | 68 |  |
| Birmingham, AL (33°30′ N) | 43 | 63 | 80 | 63 |  |
| Charleston, SC (33° N) | 48 | 64 | 80 | 66 |  |

1. Label the Pacific Ocean and the Atlantic Ocean on the map.

2. Draw a line of arrows on the map to show the direction of the prevailing winds.

3. Fill in the range of average temperatures for each city on the chart.

4. Which city has the smallest temperature range? _____

    Why? _____

5. Which city has the next smallest temperature range? _____

    Why? _____

6. How do the summer and winter temperatures of an inland city differ from those of a coastal city with the ocean on its windward side? _____

    _____

    Why? _____

120   GEOGRAPHY AND CLIMATE

# TEMPERATURE HIGHS AND LOWS AROUND THE WORLD

## Record Highs

| Location (Latitude) | Record High Temperature (°F) | Record for | Date |
|---|---|---|---|
| 1. El Azizia, Libya (32° N 13° E) | 136 | The world | Sept. 13, 1922 |
| 2. Death Valley, California (36° N 116° W) | 134 | Western Hemisphere | July 10, 1913 |
| 3. Tirat Tsvi, Israel (32° N 35° E) | 129 | Asia | June 21, 1942 |
| 4. Cloncurry, Queensland (21° S 140° E) | 128 | Australia | Jan. 16, 1889 |
| 5. Seville, Spain (37° N 6° W) | 122 | Europe | Aug. 4, 1881 |
| 6. Rivadavia, Argentina (35° S 65° W) | 120 | South America | Dec. 11, 1905 |
| 7. Midale, Saskatchewan (49° N 103° W) | 113 | Canada | July 5, 1937 |
| 8. Fort Yukon, Alaska (66° N 145° W) | 100 | Alaska | June 27, 1915 |
| 9. Pahala, Hawaii (19° N 155° W) | 100 | Hawaii | April 27, 1931 |
| 10. Esparanza, Antarctica (63° S 65° W) | 58 | Antarctica | Oct. 20, 1956 |

## Record Lows

| Location (Latitude) | Record Low Temperature (°F) | Record for | Date |
|---|---|---|---|
| 1. Vostok, Antarctica (72° S 12° E) | −127 | The world | Aug. 24, 1960 |
| 2. Verkhoyansk, USSR (67° N 133° E) | −90 | Northern Hemisphere | Feb. 7, 1892 |
| 3. Northice, Greenland (72° N 45° W) | −87 | Greenland | Jan. 9, 1954 |
| 4. Snag, Yukon (62° N 140° W) | −81 | North America | Feb. 3, 1947 |
| 5. Prospect Creek, Alaska (66° N 152° W) | −80 | Alaska | Jan. 23, 1971 |
| 6. Rogers Pass, Montana (47° N 113° W) | −70 | U.S. (excluding Alaska) | Jan. 20, 1954 |
| 7. Sarmiento, Argentina (34° S 69° W) | −27 | South America | June 1, 1907 |
| 8. Ifrane, Morocco (33° N 5° W) | −11 | Africa | Feb. 11, 1935 |
| 9. Charlotte Pass, Australia (36° S 148° E) | −8 | Australia | July 22, 1949 |
| 10. Mt. Haleakala, Hawaii (20° N 156° W) | 14 | Hawaii | Jan. 2, 1961 |

*The Weather Report* © 1989

NAME _____

# TEMPERATURE HIGHS AND LOWS AROUND THE WORLD: QUESTIONS

Refer to a map of the world to answer the following questions.

1. Find Libya (which holds the world's high temperature record) on the map. Find Antarctica (which holds the world's low temperature record) on the map.

   Which one is closer to the equator? _____

2. What is the record high temperature in Alaska? _____

   What is the record low temperature in Alaska? _____

   What is the range of temperature—from record high to record low—in Alaska?

   _____

3. What is the record high temperature in Hawaii? _____

   What is the record low temperature in Hawaii? _____

   What is the range of temperature—from record high to record low—in Hawaii?

   _____

4. Does Hawaii or Alaska have a greater temperature range? _____

   Give two reasons why Hawaii and Alaska have different temperature ranges.

   _____

   _____

   _____

   _____

*The Weather Report © 1989*

122    GEOGRAPHY AND CLIMATE

# AVERAGE HUMIDITIES BY LATITUDE

*The Weather Report © 1989*

GEOGRAPHY AND CLIMATE    123

# PRECIPITATION HIGHS AND LOWS AROUND THE WORLD

## Record Rainfalls

| Location | Amount (inches) | Record for | Date |
|---|---|---|---|
| 1. Cherrapunji, India (25° N 91° E) | 1,042 | 1 year (world) | Aug. 1860–Aug. 1861 |
| 2. Mt. Waialeale, Kauai, Hawaii (22° N 159° W) | 460 | Annual average (world) | Annual Average |
| 3. Cherrapunji, India (25° N 91° E) | 366 | 1 month (world) | July 1861 |
| 4. Belouve, Reunion Island | 53 | 12 hours (world) | Feb. 28, 1964 |
| 5. Alvin, Texas (29° N 95° W) | 43 | 24 hours (U.S.) | July 25, 1979 |
| 6. Holt, Missouri (30° N 30° W) | 12 | 42 minutes (world) | June 22, 1947 |
| 7. Unionville, Maryland (15° N 30° W) | 1.2 | 1 minute (world) | July 4, 1956 |

## Record Snowfalls

| Location | Amount (inches) | Record for | Date |
|---|---|---|---|
| 8. Paradise Ranger Station, Mt. Rainier, Washington (47° N 121° W) | 1,122 | 1 year (U.S.) | 1971–1972 |
| 9. Tamarack, California (38° N 119° W) | 390 | 1 month (world) | January 1911 |
| 10. Mt. Shasta Ski Bowl, California (41° N 122° W) | 189 | 1 snowstorm (world) | Feb. 13–19, 1959 |
| 11. Silverlake, Boulder Co., Colorado (40° N 105° W) | 76 | 24 hours (world) | April 14–15, 1921 |

## Record Dry Spots

| Location | Amount (inches) | Record for | Date |
|---|---|---|---|
| 12. Arica, Chile (18° N 70° W) | 0.03 | Lowest annual average rainfall (world) | (annual average) |
| 13. Bagdad, California (35° N 116° W) | 0.0 | Longest period without measurable precipitation (U.S.) | Aug. 1909 to May 1912 (933 days) |

*The Weather Report © 1989*

NAME _____

# PRECIPITATION HIGHS AND LOWS AROUND THE WORLD: QUESTIONS

Refer to a world map for help in answering the following questions.

1. Give two reasons why Hawaii and Reunion Island are the locations of rainfall records. _____

2. In what area of the United States have the record snowfalls occurred?

   _____

   Give two reasons for heavy snows in this area. _____

   _____

3. Locate the Sahara Desert. Give two reasons why there is so little precipitation there. _____

   _____

4. Look up *monsoon* in a dictionary or encyclopedia. Then explain why Cherrapunji, India, holds world records for rainfall. _____

   _____

   _____

   _____

   _____

   _____

*The Weather Report* © 1989

GEOGRAPHY AND CLIMATE     125

NAME _____

# Geography and Climate Review Test

1. Which two of the following are true about temperatures as you go farther from the equator?

   A. Warmer overall
   B. Cooler overall
   C. Less temperature change between seasons
   D. More temperature change between seasons

2. Which of the following is not a reason for an area receiving large amounts of precipitation?

   A. Being on the western slope of a mountain
   B. Lying in a typical storm track
   C. Being near the equator
   D. Being near a large body of water

3. Look at the following map. Which of the three cities shown—Nairobi (1.17° S), Mexico City (19.24° N), and New York City (40.43° N)—would you expect to be the warmest overall? _____

   Which one would you expect to be coolest overall? _____

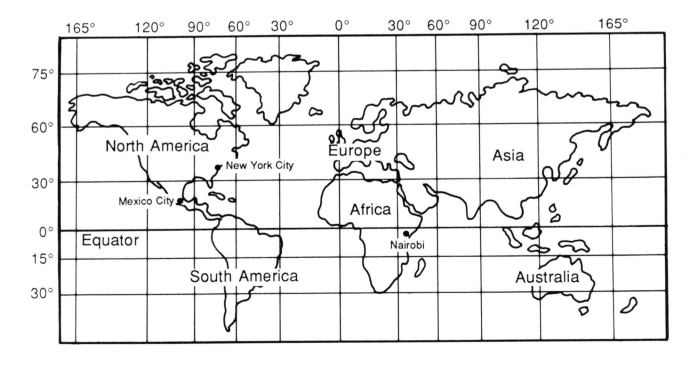

126  GEOGRAPHY AND CLIMATE

# Geography and Climate Review Test continued

4. Look at the following map. Which of the four cities shown—San Francisco, Reno, Denver, and Kansas City—would you expect to be coolest in the summer and warmest in the winter? _____

GEOGRAPHY AND CLIMATE 127

# CHAPTER 11
# RECORDING AND FORECASTING WEATHER

Weather changes from day to day and area to area. Even though over 10,000 land-based weather stations are in operation around the world, using data from computers and weather satellites, weather reports for specific localities are not always accurate, because of variations caused by terrain. Many times a *local* weather station can make a more accurate prediction for the weather in its area.

Your class can make quite reliable predictions for local weather conditions by recording the data about temperature, cloud types, precipitation, wind speed and direction, air pressure, and humidity. But perhaps the most important information to help you predict your local weather is knowing generally what to expect in your area during each time of year.

## Concept

Weather can be predicted from information about changes in temperature, cloud types, precipitation, humidity, wind speed and direction, and air pressure in combination with records of average weather conditions for a given location.

## Activities

1. **Charting weather: early primary grades.** Use a felt board to show the daily weather report. Cut thin strips of felt for a weekly calendar that looks like this:

| January 25-29 | | | | | |
|---|---|---|---|---|---|
| | Mon. | Tues. | Wed. | Thurs. | Fri. |
| Temp. | | | | | |
| Wind | | | | | |
| Clouds | | | | | |
| Precip. | | | | | |

Make labels for the school months (January, February, March, April, May, June, September, October, November, December), for the days of the week (Monday through Friday), for dates (1 through 31), and for weather conditions (Temperature, Wind, Clouds, and Precipitation). Use the symbols on page 131 as patterns. Trace or copy the symbols on appropriate colors of felt scrap. You'll need five copies of the symbols for weather conditions that are common in your area; one or two copies will be enough for the less common conditions.

Design other weather symbols for other weather conditions that may occur in your area, such as drizzle, sleet, tornadoes, hurricanes, smog.

Assign one student to give the weather report to the class each day by describing the weather conditions and putting the appropriate symbols in the boxes.

2. **Charting weather: upper grades.** Duplicate "Sample Weather Chart" (page 132) for your students. As you go over the kinds of data in the chart, use the

129

opportunity to review the kind of instrument used to record each kind of data.

Before they fill in the predictions for the last three days on the chart, hand out "Instant Weather Forecast Chart" (page 133). This chart is based on general patterns and is most applicable to the eastern two-thirds of the United States and works best during the fall, winter, and spring, when weather systems are most active.

Have your students refer to the chart to make the predictions for the last three days in the sample weather chart. After your students have attempted their own predictions, discuss the responses as a class.

**3. Recording and predicting weather.** Duplicate "Weekly Weather Chart" to use to record the data collected from your class weather station every week.

The recording could be the responsibility of a different student each day or of the whole class. Have your students compare their records and predictions with those that are published in local newspapers and with those that are broadcast on national network weather reports.

**4. Making a weather map.** Hand out the worksheet "Weather Map"(p.135). If you obtained satellite photos for the Fronts and Storms chapter, have your students diagram it on the map. If you do not have a satellite photo, give them an imaginary description of U.S. weather conditions (or use the conditions described in the newspaper or a videotape of a satellite weather photo) to diagram on the map.

**5. Predicting weather from weather maps.** Have your students bring in the weather maps from the newspaper (and, if possible, tape the satellite pictures from the TV weather report in order to show the actual movement of the weather systems.) Trace the daily movement of storms, changes in temperatures, and movements of high- and low-pressure systems. This can be done as a daily weather report. A large, laminated U.S. weather map is ideal for this purpose. How long does it take the various weather systems to move? In what general directions do they move? Where do you predict they will be tomorrow? Next week?

## ANSWER KEY:
## Sample Weather Chart

**1.** Falling air pressure, increasing clouds that are lowering and thickening, shift in wind direction, higher humidity.
**2.** November 18 or 19. **3.** Nov. 19: clearing; Nov. 20: fair; Nov. 21: fair, with a possible change in 24–48 hours.

# WEATHER CHART PATTERNS

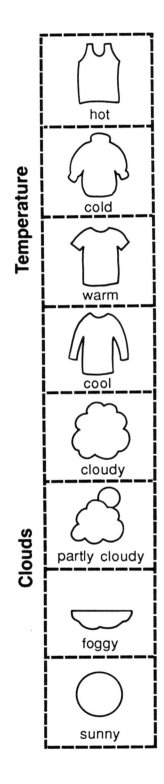
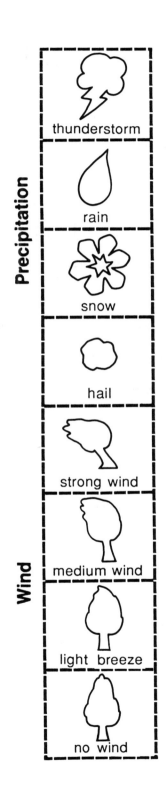

RECORDING AND FORECASTING WEATHER   131

NAME _____

# SAMPLE WEATHER CHART

Dates

| | Nov. 13 | Nov. 14 | Nov. 15 | Nov. 16 | Nov. 17 | Nov. 18 | Nov. 19 | Nov. 20 | Nov. 21 |
|---|---|---|---|---|---|---|---|---|---|
| High temperature (°F) | 75° | 76° | 84° | 76° | 72° | 64° | 66° | 68° | 69° |
| Low temperature (°F) | 38° | 41° | 43° | 40° | 42° | 46° | 43° | 36° | 34° |
| Current temperature (°F) | 72° | 74° | 78° | 71° | 68° | 60° | 61° | 65° | 63° |
| Cloud type | cirrus | none | none | cirro-stratus | alto cum., stratus | stratus | cumulus | none | cirrus |
| % cloud cover | 15% | none | none | 25% | 40% | 90% | 50% | none | 10% |
| Air Pressure | 29.98 | 30.11 | 30.09 | 29.96 | 29.89 | 29.82 | 29.98 | 30.06 | 30.03 |
| Humidity | 46% | 28% | 22% | 44% | 47% | 75% | 61% | 46% | 44% |
| Wind direction | NW | N | N | SW | SW | S-SW | W | NW | NW |
| Wind speed (mph) | 2–6 | 3–9 | 6–12 | 3–7 | 3–6 | 5–10 | 4–8 | 5–10 | 3–6 |
| Rain daily (inches) | none | none | none | none | none | .26 | .13 | none | none |
| Rain total (inches) | 1.38 | 1.38 | 1.38 | 1.38 | 1.38 | 1.64 | 1.77 | 1.77 | 1.77 |
| Prediction | fair | fair | fair | chng. | rain in 12 hours | rain | | | |

1. What were the signs that rain was approaching?

   _____

   _____

2. On what day did the front pass through? _____

3. Fill in the predictions for November 19, 20, and 21 on the chart.

*The Weather Report* © 1989

# INSTANT WEATHER FORECAST CHART

| Sea-Level Pressure | Pressure Tendency | Surface-Wind Direction | Sky Condition | 24-Hr. Forecast (see code below) |
|---|---|---|---|---|
| 30.20 or higher | rising, steady or falling | any direction | clear, high clouds, Cu | 1, 18 (in winter, 14) |
| 30.19 to 30.00 | rising or steady | SW, W, NW, N | clear, high clouds or Cu | 1, 18 |
| | falling | SW, S, SE | clear, high clouds | 1, 3, 17, 5 |
| | | | middle or low clouds | 6, 17 |
| | | E, NE | middle or low clouds | 6, 14 |
| | | | clear or high clouds | 3, 5, 14 |
| 29.99 to 29.80 | rising | SW, W, NW, N | clear | 1, 14 |
| | | | overcast | 2, 16 |
| | | | precipitation | 11, 2, 16 |
| | falling | any direction | clear | 3, 17 (dry climate summer, 1, 15) |
| | | SW, S, SE | high clouds | 3, 17, 5 |
| | | | middle or low clouds | 7 |
| | | E, NE | middle or low clouds | 7, 12, 14 |
| | | SE, E, NE | overcast, precipitation | 9 |
| | | S, SE | overcast, precipitation | 10, 13 |
| 29.79 or below | rising | SW, W, NW, N | clear | 1, 12 |
| | | | overcast | 2, 12, 16 |
| | | | overcast w/precipitation | 11, 12, 16 |
| | | NE | overcast | 4, 12, 13, 14 |
| | | | overcast w/precipitation | 11, 12, 13, 14 |
| | | SW, S, SE | clear | 3, 6, 8, 12, 15 |
| | | | overcast | 7, 8, 12, 13 |
| | falling | SW, S, SE | overcast w/precipitation | 8, 10, 12, 13, 16 |
| | | N | overcast | 4, 14 |
| | | E, NE | overcast | 7, 12, 14 |
| | | | overcast w/precipitation | 8, 9, 12, 13 |

## Weather Forecast Code

1 = clear or scattered clouds
2 = clearing
3 = increasing clouds
4 = continued overcast
5 = precipitation possible within 24 hours
6 = precipitation possible within 12 hours
7 = precipitation possible within 8 hours
8 = possible period of heavy precipitation
9 = precipitation continuing

10 = precipitation ending within 12 hours
11 = precipitation ending within 6 hours
12 = windy
13 = possible wind shift to W, NW, or N
14 = continued cool or cold
15 = continued mild or warm
16 = turning colder
17 = slowly rising temperatures
18 = little temperature change

*The Weather Report © 1989*

RECORDING AND FORECASTING WEATHER    133

NAME _____

# WEEKLY WEATHER CHART

**Dates**

| | | | | | | | |
|---|---|---|---|---|---|---|---|
| High temperature (°F) | | | | | | | |
| Low temperature (°F) | | | | | | | |
| Current temperature (°F) | | | | | | | |
| Cloud type | | | | | | | |
| % cloud cover | | | | | | | |
| Air Pressure | | | | | | | |
| Humidity | | | | | | | |
| Wind direction | | | | | | | |
| Wind speed (mph) | | | | | | | |
| Rain daily (inches) | | | | | | | |
| Rain total (inches) | | | | | | | |
| Prediction | | | | | | | |

*The Weather Report* © 1989

NAME _____

# WEATHER MAP

- ▲▲ cold front (peaks on side of line toward which front is advancing)
- ⌒⌒ warm front (bumps on side of line toward which front is advancing)
- occluded front
- stationary front
- ➤ wind direction

- ∴ snow
- ////// rain
- Ⓗ high pressure area
- Ⓛ low pressure area
- ○ clear
- ◐ partly cloudy
- ● cloudy

*The Weather Report* © 1989

RECORDING AND FORECASTING WEATHER    135

# CHAPTER 12
# BUILDING YOUR OWN WEATHER STATION

Your class's weather station can be plain or fancy, depending on your time and budget. On the following pages you will find instructions for building a simple instrument shelter and instruments. Many weather observation instruments can be inexpensively purchased at hardware stores or through scientific supply catalogues (see "Science Equipment Suppliers," page 151).

To observe and predict the weather, your weather station should be equipped with the following instruments:

1. a maximum-minimum thermometer (remember to reset it every day and to record the high as late in the afternoon as possible),
2. a cloud chart (see page 39 for a simple cloud chart),
3. a barometer (see pages 138 and 139 for instructions for simple barometers),
4. a hygrometer (see page 140),
5. a wind vane or windsock (see pages 141 and 142, or use the school flag to observe wind direction),
6. an anemometer (see page 143, or use the Beaufort Wind Scale on page 63 for a rough estimate of wind speed), and
7. a rain gauge (you can use a coffee can, as described on page 46).

# Soap-Bottle Barometer

## Materials

Empty plastic soap bottle (with cap)

Small block of wood

Broom straw

Index card

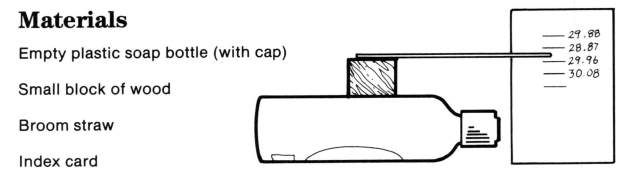

## Directions

1. Squeeze all the air out of the soap bottle. Screw on the cap and seal it tightly with glue.
2. Lay the bottle horizontally and glue the block of wood on top of it.
3. Glue one end of the broom straw to the wood block.
4. Mount the index card behind the straw.
5. To calibrate the fluctuations of the straw to air pressure, you'll need a real barometer. Mark the straw's fluctuations on the card and label the marks according to the barometer readings. Once several pressures are recorded and calibrated, the soap bottle barometer will be all you need to measure the air pressure.

NOTE: This barometer will work only if it is used at a specific temperature each day. The barometer is sensitive to both pressure and temperature, so the variable of temperature must be neutralized.

# Tin-Can Barometer

## Materials

Coffee can

Plastic wrap or bulb of large balloon

Straw

Index card

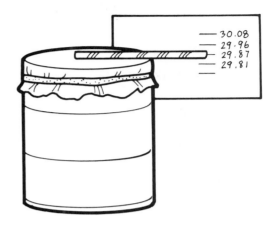

## Directions

1. Cover the top of the coffee can with plastic wrap or the bulb of a balloon. (A balloon is much more likely to stay airtight over several days.) Hold the plastic wrap or balloon in place with a rubber band. The cover should be taut and airtight.

2. Place the straw so two thirds of it is on the can. Tape the straw to the middle of the plastic wrap or balloon.

3. Tape the index card to the can behind the straw.

4. Record the movements of the straw on the index card. Calibrate these fluctuations with a real barometer. High air pressure will make the plastic wrap or balloon cave in and the straw go up. Low air pressure will make the plastic wrap or balloon puff up and the straw go down.

NOTE: Like the soap bottle barometer, the tin can barometer must be used at the same temperature every day, because it is sensitive to changes in temperature as well as pressure.

**BUILDING YOUR OWN WEATHER STATION**

# Milk-Carton Hygrometer

## Materials

Empty 1-quart milk carton

Two thermometers

Strip of absorbent cloth

String or rubber bands

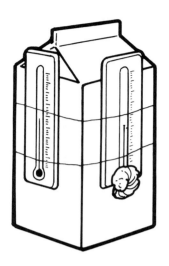

## Directions

1. Cut a hole about 2½ inches from the bottom of the milk carton.
2. Fill the carton with water up to the hole.
3. Fasten the two thermometers to the milk carton with string or with rubber bands. One thermometer should have its bulb just above the hole.
4. Wrap the bulb of the thermometer above the hole with the strip of cloth. Push the tail of the cloth through the hole.
5. To determine relative humidity, refer to the table on page 29.

# Wind Vane

## Materials

Square 1" × 4" piece of wood

Plastic straw

Headless nail

5" long piece of cardboard

Aluminum foil

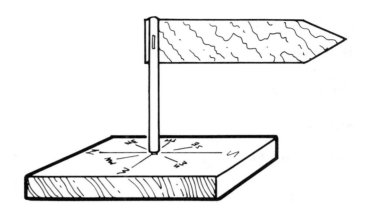

## Directions

1. Mark and label the cardinal and intermediate directions on the piece of wood.
2. Hammer the headless nail into the center of the wood just until it holds firm.
3. Cut the plastic straw so it is one-third longer than the headless nail.
4. Cut the piece of cardboard into an arrow. Cover it with aluminum foil (to prevent rotting).
5. Staple the cardboard arrow to the top of the straw. Put the straw on the headless nail.
6. Place your wind vane outside, making sure that North on the base is pointing north.

BUILDING YOUR OWN WEATHER STATION

# Windsock

## Materials

Heavy cloth, about 36" × 24"

4 lengths (about 10" each) of heavy wire

Wire coat hanger

Stick, about 36" long

Large nail

Wooden spool

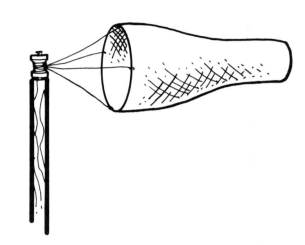

## Directions

1. Form the hanger into a loop about 9 inches in diameter. Attach the 4 wires to the loop at 4 equidistant points.

2. Cut cloth into a sleeve in the shape shown here. Sew sides together, making a cone, and sew wide end of cone to loop.

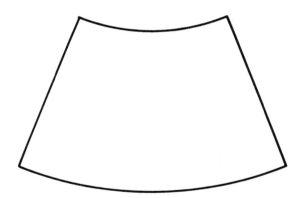

3. Twist exposed ends of wires around the spool.

4. Place the nail through the spool, and hammer nail into end of stick so that the spool can pivot freely around nail.

5. Place the windsock outdoors. Nail the stick to a tall post or to a rooftop away from obstructions so the sock can swing freely in the wind.

6. Observe the position of the sock at various times. The large end of the sock will catch the wind; the small end will point away from the direction from which the wind is blowing, or it will droop if there is not enough wind to keep it extended.

# Anemometer

## Materials

Square 1" × 6" piece of wood

Plastic straw

Two nails, one headless and one with a head

3 small paper cups

1 aluminum pie plate

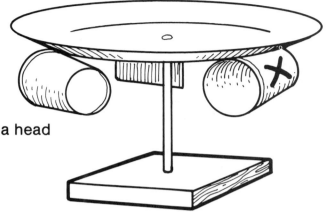

## Directions

1. Hammer the headless nail into the wood, leaving at least ½ inch protruding.
2. Put the plastic straw onto the nail.
3. Staple the three cups horizontally and facing in the same direction an equal distance apart at the edge of the underside of the plate. Mark one of the cups with a big X.
4. Center the plate on the straw and balance it. Put the headed nail through the plate into the straw.
5. To calibrate wind speed, you can take the anemometer in a car on a calm day and count the revolutions while the car is being driven at 3, 5, 10, 15, 20 mph (up to whatever speed the wind usually reaches in your area).
6. Place the anemometer outside and correlate the revolutions with those you counted in the car. This will give you the wind speed.

# A Simple Instrument Shelter

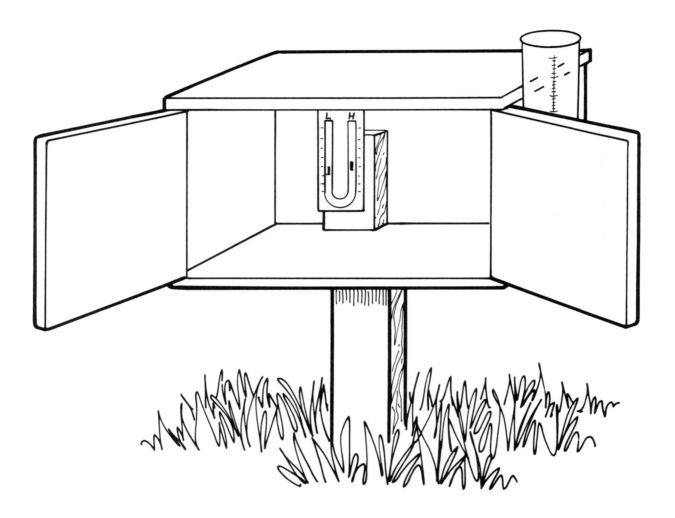

## Some Tips for Setting Up an Instrument Shelter

1. The opening should face north, so the sun will never hit the thermometer.
2. The shelter should be painted white, so it will reflect the sun.
3. The maximum-minimum thermometer should be mounted on a small block of wood so that it is not directly touching the back of the shelter.
4. The opening of the rain gauge should be above the rest of the shelter.
5. The shelter should be in an open area and tall enough so the children can read it at eye level.

Instructions for building an instrument shelter are available from National Weather Service stations.

NAME _____

# Review Test for Weather Unit

Use the following words to fill in the blanks in Sentences 1–12.

observing     humidity     stratus     frost     predicting     cirrus     dew

atmosphere     hygrometer     cumulus     meteorologists     precipitation

1. When you find out about weather by seeing, feeling, and listening, you are

   _____ .

2. When you use observations to say what the weather may be in the future,

   you are _____ .

3. People whose job is to predict and record the weather are called

   _____

4. The "blanket" of air that surrounds this planet is called the _____ .

5. When a lot of water vapor is in the air, the _____ is high.

6. Water vapor in the air is measured with a _____ .

7. Rain, snow, sleet, and hail are all different kinds of _____ .

8. _____ clouds are high, thin, and made of bits of ice; they
   indicate a change of weather.

9. _____ clouds are dark and low; they indicate rain.

10. _____ clouds are puffy; they indicate fair weather.

11. Tiny drops of water that form on the ground are called _____ .

12. If the temperature is below freezing, dew becomes _____ .

*The Weather Report © 1989*

# Review Test for Weather Unit continued _____

13. Put the letter of the kind of weather information in the blank by the instrument that measures it.

_____ Hygrometer

_____ Barometer

_____ Rain gauge

_____ Wind vane

_____ Anemometer

_____ Thermometer

A. Temperature

B. Wind direction

C. Humidity

D. Rainfall

E. Air pressure

F. Wind speed

14. Give a prediction for each set of weather data.

Cloud type: cumulus

Air pressure: 29.98 and rising

Wind direction: NW

A. Prediction: _____

Cloud type: altocumulus, cirrus

Air pressure: 30.01 and falling

Wind direction: SW

B. Prediction: _____

15. Put the letter of the kind of front by the symbol that represents it.

A. Warm front

B. Occluded front

C. Stationary front

D. Cold front

*The Weather Report* © 1989

146    REVIEW TEST

NAME _____

# Crossword Puzzle

**Across**

3. The study of weather
6. A product of sun and rain
8. Moisture from the sky
13. Moisture level in the air
15. Polluted air
16. What rain may turn into when it is cold
18. High, thin clouds
19. Where warm air and cold air meet
20. What a thermometer measures

**Down**

1. A severe dry period
2. Frozen drops of water
4. The smallest amount of rainfall
5. Forecast
7. Ground level clouds
9. Thunderclouds
10. What is found on cold, moist mornings
11. Fair weather clouds
12. Instrument to measure air pressure
14. The most violent kind of storm that occurs over land
17. The coldest season

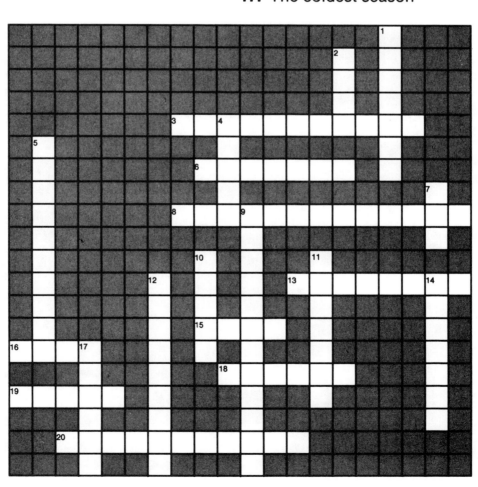

REVIEW TEST 147

ANSWER KEY:
## Review Test for Weather Unit

1. observing.  2. predicting.
3. meteorologists.  4. atmosphere.
5. humidity.  6. hygrometer.
7. precipitation.  8. cirrus.
9. stratus.  10. cumulus.
11. dew.  12. frost. 13. Hygrometer: C;
Barometer: E; Rain gauge: D; Wind vane: B;
Anemometer: F; Thermometer: A. 14. A:
Fair; B: Increasing chance of rain or
snow.  15. 〰 : D; 〰: A; 〰: B;
〰 : C.

ANSWER KEY:
## Crossword Puzzle

*The Weather Report* © 1989

148    REVIEW TEST

# GLOSSARY

**air pressure**   The pressure or weight of the air over a given point.

**altitude**   The elevation above or below sea level.

**altocumulus**   Middle clouds, usually white or gray. Often occur in layers or patches.

**anemometer**   An instrument that measures wind speed.

**atmosphere**   All of the air surrounding the earth.

**barometer**   An instrument that measures air pressure.

**cirrus**   A high cloud made of ice crystals. Often thin, white, and in narrow bands.

**cold front**   Caused by a cold air mass wedging under a warm air mass.

**condensation**   The process of water vapor becoming a liquid.

**convections**   Atmospheric motions that result in warm air rising and cool air sinking.

**cumulonimbus**   A dense and vertically developed cloud, often with an anvil shape at the top. Frequently brings thunderstorms and hail.

**cumulus**   A flat-based cloud with a bulging upper part that may resemble a cauliflower. Usually brings fair weather, but if it exhibits extreme vertical growth, showers can occur.

**current**   The flow of water or air in a definite direction.

**dew**   Water that has condensed onto objects near the ground when their temperatures have fallen below the dew point of the air.

**dew point**   The temperature to which the air must be cooled for saturation to occur.

**evaporation**   The process of liquid changing into a gas or vapor.

**five-hundred-millibar graph**   A chart showing the air pressure patterns at 20,000 feet above the earth's surface.

**forecast**   To predict the weather.

**front**   The meeting of two distinct air masses.

**frost**   Ice covering resulting from condensation when the air temperature is below freezing and it reaches its dew point.

**hail**   Particles of ice, usually falling from a cumulonimbus cloud, that range from the size of a pea to the size of a golf ball.

**heiligenschein**   A faint white ring surrounding the shadow of an observer's head on a dew-covered lawn.

**humidity**   A measurement of water vapor in the air.

**hurricane**   A severe tropical storm containing winds above 74 miles per hour.

**hygrometer**   An instrument for measuring the air's water vapor content, or humidity.

**latitude**   Distance north or south from the equator, measured in degrees.

**lightning rod**   A metal rod placed high on a building and grounded to divert lightning from the structure.

**mammatus cloud**   A cloud that looks like pouches hanging from the bottom of another cloud.

**marine influence**   The effect of ocean temperature on the weather of nearby land.

**maximum-minimum thermometer**   A device that automatically measures the daily high and low temperatures.

**meteorologist**   A person who studies the atmosphere and its interaction with the earth's surface.

**microclimate**   The climate or weather within a small space or area.

**precipitation** Any form of water—liquid or solid—that falls from the atmosphere and reaches the ground.

**prediction** Calculations about what the weather will be.

**rain gauge** A device for measuring rainfall amounts.

**relative humidity** The ratio of the amount of water vapor in the air compared to the amount of water vapor that the air can hold at that temperature.

**satellite picture** A photograph of the earth from a satellite in orbit. It can show cloud, wind, and precipitation patterns.

**sleet** A type of precipitation that is small ice pellets. Usually occurs when rain freezes on its way to the ground.

**stationary front** A front that is not distinctly moving in any direction.

**stratus** A low, gray cloud layer that often brings drizzle.

**thermometer** An instrument for measuring temperature.

**topography** The surface features of any region.

**tornado** An intense, rotating funnel or rope of air that descends from a cumulonimbus cloud and touches the ground.

**warm front** A front where warm air overrides and replaces cold air.

**wind chill** The cooling effect of a combination of temperature and wind. Loss of body heat is the result.

**wind vane** An instrument for measuring wind direction.

# RESOURCES

The following organizations provide free or inexpensive weather-related materials for the classroom teacher.

American Meteorological Society
45 Beacon Street
Boston, MA 02108
(free pamphlets and climate information)

Association of American Weather Observers (AAWO)
401 Whitney Blvd.
Belvidere, IL 61008
(monthly weather newsletter provided to members, who contribute data from all over the country; newsletter also includes other related weather information)

Mindscape
3444 Dundee Road
Northbrook, IL 50052
1-800-221-9884
(excellent weather forecast computer software that complements the daily weather charting in this book)

National Climate Data Center
Federal Building
Asheville, NC 29901-2696
(free pamphlets and climate information, including publications by the National Oceanic and Atmospheric Administration and the National Weather Service)

*WeatherWise*
Heldref Publications
4000 Albemarle Street, NW
Washington, DC 20016
(up-to-date professional magazine on weather trends, patterns, and information)

# SCIENCE EQUIPMENT SUPPLIERS

The following science supply companies stock weather instruments for home and school. Write for their catalogs.

American Weather Enterprises
P.O. Box 1383
Media, PA 19063

Carolina Biological Supply Company
2700 York Rd.
Burlington, NC 27215-3398
(toll free 1-800-334-5551)

Cloud Chart, Inc.
P.O. Box 21298
Charleston, SC 29413
803-577-5268

Edmund Scientific
101 E. Gloucester Pike
Barrington, NJ 08007
609-547-3488

Fisher Scientific
4901 West LeMoyne St.
Chicago, IL 60651
(toll free 1-800-621-4769)

Frey Scientific
905 Hickory Lane
Mansfield, OH 44905
(toll free 1-800-225-FREY)

Hubbard Scientific
P.O. Box 104
Northbrook, IL 60065
(toll free 1-800-323-8368)

Nasco Science
1524 Princeton Avenue
Modesto, CA 95352
(toll free 1-800-558-9595)

Simerl Instruments
238 West Street
Annapolis, MD 21401
301-849-8667

Wind and Weather
P.O. Box 2320
Mendocino, CA 95460
707-937-0323

# APPENDIX: Climatic Data for Selected U.S. Cities—Maximum and Minimum Temperatures and Precipitation

The data in this chart is from the National Weather Service. It shows the normal temperatures (in Fahrenheit) and the normal precipitation levels (in inches) over the 30-year period from 1951 to 1980.

**Abbreviations:**  co = city office weather station      ap = airport weather station

| | JAN | FEB | MAR | APR | MAY | JUN | JUL | AUG | SEP | OCT | NOV | DEC | ANN'L |
|---|---|---|---|---|---|---|---|---|---|---|---|---|---|
| **Birmingham (co), AL** | | | | | | | | | | | | | |
| Maximum Temperatures | 51.7 | 56.6 | 64.8 | 75.0 | 81.7 | 88.2 | 90.6 | 89.8 | 84.1 | 73.5 | 62.3 | 54.5 | 72.7 |
| Minimum Temperatures | 33.0 | 35.2 | 42.2 | 50.5 | 58.4 | 66.0 | 70.0 | 69.4 | 64.0 | 50.6 | 50.7 | 35.4 | 51.3 |
| Precipitation | 4.97 | 4.64 | 6.55 | 5.30 | 3.80 | 3.17 | 4.29 | 3.88 | 4.55 | 2.77 | 3.51 | 4.83 | 52.16 |
| **Anchorage, AK** | | | | | | | | | | | | | |
| Maximum Temperatures | 20.0 | 25.5 | 31.7 | 42.6 | 54.2 | 61.8 | 65.1 | 63.2 | 55.2 | 40.8 | 27.9 | 20.4 | 42.4 |
| Minimum Temperatures | 6.0 | 10.3 | 15.7 | 28.2 | 38.3 | 47.0 | 51.1 | 49.2 | 41.1 | 28.4 | 15.4 | 7.1 | 28.2 |
| Precipitation | 0.80 | 0.93 | 0.69 | 0.66 | 0.57 | 1.08 | 1.97 | 2.11 | 2.45 | 1.73 | 1.11 | 1.10 | 15.20 |
| **Juneau, AK** | | | | | | | | | | | | | |
| Maximum Temperatures | 27.4 | 33.7 | 37.4 | 46.8 | 54.7 | 61.1 | 64.0 | 62.6 | 55.9 | 47.0 | 37.5 | 31.5 | 46.6 |
| Minimum Temperatures | 16.1 | 21.9 | 25.0 | 31.3 | 38.1 | 44.2 | 47.4 | 46.6 | 42.3 | 36.5 | 28.0 | 22.1 | 33.3 |
| Precipitation | 3.69 | 3.74 | 3.34 | 2.92 | 3.41 | 2.98 | 4.13 | 5.02 | 6.40 | 7.71 | 5.15 | 4.66 | 53.15 |
| **Flagstaff, AZ** | | | | | | | | | | | | | |
| Maximum Temperatures | 41.7 | 44.5 | 48.6 | 57.1 | 66.7 | 77.6 | 81.9 | 78.9 | 74.1 | 63.7 | 51.0 | 43.6 | 60.8 |
| Minimum Temperatures | 14.7 | 16.9 | 20.4 | 25.9 | 32.9 | 40.9 | 50.3 | 48.7 | 40.9 | 30.6 | 21.5 | 15.9 | 30.0 |
| Precipitation | 2.10 | 1.95 | 2.13 | 1.35 | 0.75 | 0.57 | 2.47 | 2.62 | 1.47 | 1.54 | 1.65 | 2.26 | 20.86 |

**Phoenix, AZ**

| | | | | | | | | | | | | | |
|---|---|---|---|---|---|---|---|---|---|---|---|---|---|
| Maximum Temperatures | 65.2 | 69.7 | 74.5 | 83.1 | 92.4 | 102.3 | 105.0 | 102.3 | 98.2 | 87.7 | 74.3 | 66.4 | 85.1 |
| Minimum Temperatures | 39.4 | 42.5 | 46.7 | 53.0 | 61.5 | 70.6 | 79.5 | 77.5 | 70.9 | 59.1 | 46.9 | 40.2 | 57.3 |
| Precipitation | 0.73 | 0.59 | 0.81 | 0.27 | 0.14 | 0.17 | 0.74 | 1.02 | 0.64 | 0.63 | 0.54 | 0.83 | 7.11 |

**Little Rock, AR**

| | | | | | | | | | | | | | |
|---|---|---|---|---|---|---|---|---|---|---|---|---|---|
| Maximum Temperatures | 49.8 | 54.5 | 63.2 | 73.8 | 81.7 | 89.5 | 92.7 | 92.3 | 85.6 | 75.8 | 62.4 | 53.2 | 72.9 |
| Minimum Temperatures | 29.9 | 44.6 | 41.2 | 50.9 | 59.2 | 67.5 | 71.4 | 69.6 | 63.0 | 50.4 | 40.0 | 33.2 | 50.8 |
| Precipitation | 3.91 | 3.83 | 4.69 | 5.41 | 5.29 | 3.67 | 3.63 | 3.07 | 4.26 | 2.84 | 4.37 | 4.23 | 49.20 |

**Blue Canyon, CA**

| | | | | | | | | | | | | | |
|---|---|---|---|---|---|---|---|---|---|---|---|---|---|
| Maximum Temperatures | 43.5 | 44.9 | 45.3 | 51.3 | 60.3 | 69.2 | 77.7 | 76.5 | 72.4 | 62.8 | 51.4 | 46.3 | 58.5 |
| Minimum Temperatures | 30.7 | 31.3 | 31.0 | 35.2 | 42.8 | 51.0 | 58.9 | 57.3 | 53.3 | 45.5 | 37.1 | 32.7 | 42.2 |
| Precipitation | 14.11 | 9.93 | 8.96 | 5.45 | 2.70 | 0.86 | 0.30 | 0.55 | 0.97 | 3.93 | 8.41 | 11.70 | 67.87 |

**Eureka, CA**

| | | | | | | | | | | | | | |
|---|---|---|---|---|---|---|---|---|---|---|---|---|---|
| Maximum Temperatures | 53.4 | 54.6 | 54.0 | 54.7 | 57.0 | 59.1 | 60.3 | 61.3 | 62.2 | 60.3 | 57.5 | 54.5 | 57.4 |
| Minimum Temperatures | 41.3 | 42.6 | 42.5 | 44.0 | 47.3 | 50.2 | 51.9 | 52.6 | 51.5 | 48.3 | 45.2 | 42.2 | 46.6 |
| Precipitation | 6.99 | 5.20 | 5.05 | 2.91 | 1.60 | 0.56 | 0.10 | 0.37 | 0.90 | 2.71 | 5.90 | 6.22 | 38.51 |

**Los Angeles (co), CA**

| | | | | | | | | | | | | | |
|---|---|---|---|---|---|---|---|---|---|---|---|---|---|
| Maximum Temperatures | 66.6 | 68.5 | 68.7 | 70.9 | 73.2 | 77.9 | 83.8 | 84.1 | 83.0 | 78.5 | 72.7 | 68.1 | 74.7 |
| Minimum Temperatures | 47.7 | 49.2 | 50.2 | 53.0 | 56.6 | 60.4 | 64.3 | 65.3 | 63.7 | 59.2 | 52.7 | 48.4 | 55.9 |
| Precipitation | 3.69 | 2.96 | 2.35 | 1.17 | 0.23 | 0.03 | 0.00 | 0.12 | 0.27 | 0.21 | 1.85 | 1.97 | 14.85 |

**Sacramento, CA**

| | | | | | | | | | | | | | |
|---|---|---|---|---|---|---|---|---|---|---|---|---|---|
| Maximum Temperatures | 52.6 | 59.4 | 64.1 | 71.0 | 79.7 | 87.4 | 93.3 | 91.7 | 87.6 | 77.7 | 63.2 | 53.2 | 76.7 |
| Minimum Temperatures | 37.9 | 41.2 | 42.4 | 45.3 | 50.1 | 55.1 | 57.9 | 57.6 | 55.8 | 50.0 | 42.8 | 37.9 | 47.8 |
| Precipitation | 4.03 | 2.88 | 2.06 | 1.31 | 0.33 | 0.11 | 0.05 | 0.07 | 0.27 | 0.86 | 2.23 | 2.90 | 17.10 |

**San Diego, CA**

| | | | | | | | | | | | | | |
|---|---|---|---|---|---|---|---|---|---|---|---|---|---|
| Maximum Temperatures | 65.2 | 66.4 | 65.9 | 67.8 | 68.6 | 71.3 | 75.6 | 77.6 | 76.8 | 74.6 | 69.9 | 66.1 | 70.5 |
| Minimum Temperatures | 48.4 | 50.3 | 52.1 | 54.5 | 58.2 | 61.2 | 64.9 | 66.8 | 65.1 | 60.3 | 53.6 | 48.7 | 57.0 |
| Precipitation | 2.11 | 1.43 | 1.60 | 0.78 | 0.24 | 0.06 | 0.01 | 0.11 | 0.19 | 0.33 | 1.10 | 1.36 | 9.32 |

| | JAN | FEB | MAR | APR | MAY | JUN | JUL | AUG | SEP | OCT | NOV | DEC | ANNL |
|---|---|---|---|---|---|---|---|---|---|---|---|---|---|
| **San Francisco (co), CA** | | | | | | | | | | | | | |
| Maximum Temperatures | 56.1 | 59.4 | 60.0 | 61.1 | 62.5 | 64.3 | 64.0 | 65.0 | 68.9 | 68.3 | 62.9 | 56.9 | 62.5 |
| Minimum Temperatures | 46.2 | 48.4 | 48.6 | 49.2 | 50.7 | 52.5 | 53.1 | 54.2 | 55.8 | 54.8 | 51.5 | 47.2 | 51.0 |
| Precipitation | 4.48 | 2.83 | 2.58 | 1.48 | 0.35 | 0.15 | 0.04 | 0.08 | 0.24 | 1.09 | 2.49 | 3.52 | 19.33 |
| **Denver, CO** | | | | | | | | | | | | | |
| Maximum Temperatures | 43.1 | 46.9 | 51.2 | 61.0 | 70.7 | 81.6 | 88.0 | 85.8 | 77.5 | 66.8 | 52.4 | 46.1 | 64.3 |
| Minimum Temperatures | 15.9 | 20.2 | 24.7 | 33.7 | 43.6 | 52.4 | 48.7 | 57.0 | 47.7 | 36.9 | 25.1 | 18.9 | 36.2 |
| Precipitation | 0.51 | 0.69 | 1.21 | 1.81 | 2.47 | 1.58 | 1.93 | 1.53 | 1.23 | 0.98 | 0.82 | 0.55 | 15.31 |
| **Washington (National ap), DC** | | | | | | | | | | | | | |
| Maximum Temperatures | 42.9 | 45.9 | 55.0 | 67.1 | 75.9 | 84.0 | 87.9 | 86.4 | 80.1 | 68.9 | 57.4 | 46.6 | 66.5 |
| Minimum Temperatures | 27.5 | 29.0 | 36.6 | 46.2 | 56.1 | 65.0 | 69.9 | 68.7 | 62.0 | 49.7 | 39.9 | 31.2 | 48.5 |
| Precipitation | 2.76 | 2.62 | 3.46 | 2.93 | 3.48 | 3.35 | 3.88 | 4.40 | 3.22 | 2.90 | 2.82 | 3.18 | 39.00 |
| **Jacksonville, FL** | | | | | | | | | | | | | |
| Maximum Temperatures | 64.6 | 66.8 | 73.3 | 79.7 | 85.2 | 88.9 | 90.7 | 90.2 | 86.9 | 79.7 | 72.4 | 66.3 | 78.7 |
| Minimum Temperatures | 41.7 | 43.3 | 49.3 | 55.7 | 63.0 | 69.1 | 71.8 | 71.8 | 69.4 | 59.2 | 49.2 | 43.2 | 57.2 |
| Precipitation | 3.07 | 3.48 | 3.72 | 3.32 | 4.91 | 5.37 | 6.54 | 7.15 | 7.26 | 3.41 | 1.94 | 2.59 | 52.76 |
| **Miami, FL** | | | | | | | | | | | | | |
| Maximum Temperatures | 75.0 | 75.8 | 79.3 | 82.4 | 85.1 | 87.3 | 88.7 | 89.2 | 87.8 | 84.2 | 79.8 | 76.2 | 82.6 |
| Minimum Temperatures | 59.2 | 59.7 | 64.1 | 68.2 | 71.9 | 74.6 | 76.2 | 76.5 | 75.7 | 71.6 | 65.8 | 60.8 | 68.7 |
| Precipitation | 2.08 | 2.05 | 1.89 | 3.07 | 6.53 | 9.15 | 5.98 | 7.02 | 8.07 | 7.14 | 2.71 | 1.86 | 57.55 |
| **Atlanta, GA** | | | | | | | | | | | | | |
| Maximum Temperatures | 51.2 | 55.3 | 63.2 | 73.2 | 79.8 | 85.6 | 87.9 | 87.6 | 82.3 | 72.9 | 62.6 | 54.1 | 71.3 |
| Minimum Temperatures | 32.6 | 34.5 | 41.7 | 50.4 | 58.7 | 65.9 | 69.2 | 68.7 | 63.6 | 51.4 | 41.3 | 34.8 | 51.1 |
| Precipitation | 4.91 | 4.43 | 5.91 | 4.43 | 4.02 | 3.41 | 4.73 | 3.41 | 3.17 | 2.53 | 3.43 | 4.23 | 48.61 |
| **Honolulu, HI** | | | | | | | | | | | | | |
| Maximum Temperatures | 79.9 | 80.4 | 81.4 | 82.7 | 84.8 | 86.2 | 87.1 | 88.3 | 88.2 | 86.7 | 83.9 | 81.4 | 84.2 |
| Minimum Temperatures | 65.3 | 65.3 | 67.3 | 68.7 | 70.2 | 71.9 | 73.1 | 73.6 | 72.9 | 72.2 | 69.2 | 66.5 | 69.7 |
| Precipitation | 3.79 | 2.72 | 3.48 | 1.49 | 1.21 | 0.49 | 0.54 | 0.60 | 0.62 | 1.88 | 3.22 | 3.43 | 23.47 |

### Boise, ID

|  | | | | | | | | | | | | | |
|---|---|---|---|---|---|---|---|---|---|---|---|---|---|
| Maximum Temperatures | 37.1 | 44.3 | 51.8 | 60.8 | 70.8 | 79.8 | 90.6 | 87.3 | 77.6 | 64.6 | 49.0 | 39.3 | 62.8 |
| Minimum Temperatures | 22.6 | 27.9 | 30.9 | 36.4 | 44.0 | 51.8 | 58.5 | 56.7 | 48.7 | 39.1 | 30.5 | 24.6 | 39.3 |
| Precipitation | 1.64 | 1.07 | 1.03 | 1.19 | 1.21 | 0.95 | 0.26 | 0.40 | 0.58 | 0.75 | 1.29 | 1.34 | 11.71 |

### Chicago, IL

|  | | | | | | | | | | | | | |
|---|---|---|---|---|---|---|---|---|---|---|---|---|---|
| Maximum Temperatures | 29.2 | 33.9 | 44.3 | 58.8 | 70.0 | 79.4 | 83.3 | 82.1 | 75.5 | 64.1 | 48.2 | 35.0 | 58.7 |
| Minimum Temperatures | 13.6 | 18.1 | 27.6 | 38.8 | 48.1 | 57.7 | 62.7 | 61.7 | 53.9 | 42.9 | 31.4 | 20.3 | 39.7 |
| Precipitation | 1.60 | 1.31 | 2.59 | 3.66 | 3.15 | 4.08 | 3.63 | 3.53 | 3.35 | 2.28 | 2.06 | 2.10 | 33.34 |

### Des Moines, IA

|  | | | | | | | | | | | | | |
|---|---|---|---|---|---|---|---|---|---|---|---|---|---|
| Maximum Temperatures | 27.0 | 33.2 | 44.2 | 61.0 | 72.6 | 81.8 | 86.2 | 84.0 | 75.7 | 65.0 | 47.6 | 33.7 | 59.3 |
| Minimum Temperatures | 10.1 | 15.8 | 26.0 | 39.9 | 51.6 | 61.4 | 66.3 | 63.7 | 54.4 | 43.3 | 29.5 | 17.6 | 40.0 |
| Precipitation | 1.01 | 1.12 | 2.20 | 3.21 | 3.96 | 4.18 | 3.22 | 4.11 | 3.09 | 2.16 | 1.52 | 1.05 | 30.83 |

### New Orleans, LA

|  | | | | | | | | | | | | | |
|---|---|---|---|---|---|---|---|---|---|---|---|---|---|
| Maximum Temperatures | 61.8 | 64.6 | 71.2 | 78.5 | 84.5 | 89.5 | 90.7 | 90.2 | 86.8 | 79.4 | 70.1 | 64.4 | 77.7 |
| Minimum Temperatures | 43.0 | 44.8 | 51.6 | 58.8 | 65.3 | 70.9 | 73.5 | 73.1 | 70.1 | 59.0 | 49.9 | 44.8 | 58.7 |
| Precipitation | 4.97 | 5.23 | 4.73 | 4.50 | 5.07 | 4.63 | 6.73 | 6.02 | 5.87 | 2.66 | 4.06 | 5.27 | 59.74 |

### Portland, ME

|  | | | | | | | | | | | | | |
|---|---|---|---|---|---|---|---|---|---|---|---|---|---|
| Maximum Temperatures | 31.0 | 33.1 | 40.5 | 52.5 | 63.4 | 72.9 | 78.9 | 77.5 | 69.6 | 59.0 | 47.1 | 34.9 | 55.0 |
| Minimum Temperatures | 11.9 | 12.9 | 23.7 | 33.0 | 42.1 | 51.4 | 57.3 | 55.8 | 47.7 | 37.9 | 29.6 | 16.7 | 35.0 |
| Precipitation | 3.78 | 3.57 | 3.98 | 3.90 | 3.27 | 3.06 | 2.83 | 2.82 | 3.27 | 3.83 | 4.70 | 4.51 | 43.52 |

### Baltimore, MD

|  | | | | | | | | | | | | | |
|---|---|---|---|---|---|---|---|---|---|---|---|---|---|
| Maximum Temperatures | 41.0 | 43.7 | 53.1 | 65.1 | 74.2 | 82.9 | 87.1 | 85.5 | 79.1 | 67.7 | 55.9 | 45.1 | 65.0 |
| Minimum Temperatures | 24.3 | 25.7 | 33.4 | 42.9 | 52.5 | 61.5 | 66.5 | 65.7 | 58.6 | 46.1 | 36.6 | 27.9 | 45.1 |
| Precipitation | 3.00 | 2.98 | 3.72 | 3.35 | 3.44 | 3.76 | 3.89 | 4.62 | 3.46 | 3.11 | 3.11 | 3.40 | 41.84 |

### Boston, MA

|  | | | | | | | | | | | | | |
|---|---|---|---|---|---|---|---|---|---|---|---|---|---|
| Maximum Temperatures | 36.4 | 37.7 | 45.0 | 56.6 | 67.0 | 76.6 | 81.8 | 79.8 | 72.3 | 62.5 | 55.9 | 45.1 | 65.0 |
| Minimum Temperatures | 22.8 | 23.7 | 31.8 | 40.8 | 50.0 | 59.3 | 65.1 | 63.9 | 56.9 | 47.1 | 38.7 | 27.1 | 43.9 |
| Precipitation | 3.99 | 3.70 | 4.13 | 3.73 | 3.52 | 2.92 | 2.68 | 3.68 | 3.41 | 3.36 | 4.21 | 4.48 | 43.81 |

### Detroit, MI

|  | | | | | | | | | | | | | |
|---|---|---|---|---|---|---|---|---|---|---|---|---|---|
| Maximum Temperatures | 30.6 | 33.5 | 43.4 | 57.7 | 69.4 | 79.0 | 83.1 | 81.5 | 74.4 | 62.5 | 47.6 | 35.4 | 58.2 |
| Minimum Temperatures | 16.1 | 18.0 | 26.5 | 36.9 | 46.7 | 56.3 | 60.7 | 59.4 | 52.2 | 41.2 | 31.4 | 21.6 | 38.9 |
| Precipitation | 1.86 | 1.69 | 2.54 | 3.15 | 2.77 | 3.43 | 3.10 | 3.21 | 2.25 | 2.12 | 2.33 | 2.52 | 30.97 |

| | JAN | FEB | MAR | APR | MAY | JUN | JUL | AUG | SEP | OCT | NOV | DEC | ANN'L |
|---|---|---|---|---|---|---|---|---|---|---|---|---|---|
| **Duluth, MN** | | | | | | | | | | | | | |
| Maximum Temperatures | 15.5 | 21.7 | 31.9 | 47.6 | 61.3 | 70.5 | 76.4 | 73.6 | 63.6 | 53.0 | 35.2 | 21.8 | 47.7 |
| Minimum Temperatures | -2.9 | 2.2 | 13.9 | 28.9 | 39.3 | 48.2 | 54.3 | 52.8 | 44.3 | 35.4 | 21.2 | 5.8 | 28.6 |
| Precipitation | 1.20 | 0.90 | 1.78 | 2.16 | 3.15 | 3.96 | 3.96 | 4.12 | 3.26 | 2.21 | 1.69 | 1.29 | 29.68 |
| **Minneapolis-St. Paul, MN** | | | | | | | | | | | | | |
| Maximum Temperatures | 19.9 | 26.4 | 37.5 | 56.0 | 69.4 | 78.5 | 83.4 | 80.9 | 71.0 | 59.7 | 41.1 | 26.7 | 54.2 |
| Minimum Temperatures | 2.4 | 8.5 | 20.8 | 36.0 | 47.6 | 57.7 | 62.7 | 60.3 | 50.2 | 39.4 | 25.3 | 11.7 | 35.2 |
| Precipitation | 0.82 | 0.85 | 1.71 | 2.05 | 3.20 | 4.07 | 3.51 | 3.64 | 2.50 | 1.85 | 1.29 | 0.87 | 26.36 |
| **Kansas City, MO** | | | | | | | | | | | | | |
| Maximum Temperatures | 34.5 | 41.1 | 51.3 | 65.1 | 74.6 | 83.3 | 88.5 | 86.8 | 78.6 | 67.9 | 52.1 | 40.1 | 63.7 |
| Minimum Temperatures | 17.2 | 23.0 | 31.7 | 44.4 | 54.6 | 63.8 | 68.5 | 66.5 | 58.1 | 47.0 | 34.0 | 23.7 | 44.4 |
| Precipitation | 1.08 | 1.19 | 2.41 | 3.23 | 4.42 | 4.66 | 4.35 | 3.57 | 4.14 | 3.10 | 1.63 | 1.38 | 35.16 |
| **Helena, MT** | | | | | | | | | | | | | |
| Maximum Temperatures | 28.1 | 36.2 | 42.5 | 54.7 | 64.9 | 73.1 | 83.6 | 81.3 | 70.3 | 58.6 | 42.3 | 33.3 | 55.7 |
| Minimum Temperatures | 8.1 | 15.7 | 20.6 | 29.8 | 39.5 | 47.0 | 52.2 | 50.3 | 40.8 | 31.5 | 20.4 | 13.5 | 30.8 |
| Precipitation | 0.66 | 0.44 | 0.69 | 1.01 | 1.72 | 2.01 | 1.04 | 1.18 | 0.83 | 0.65 | 0.54 | 0.60 | 11.37 |
| **Omaha (Eppley ap), NE** | | | | | | | | | | | | | |
| Maximum Temperatures | 30.2 | 37.3 | 47.7 | 64.0 | 74.7 | 84.2 | 88.5 | 86.2 | 77.5 | 67.0 | 50.3 | 36.9 | 62.0 |
| Minimum Temperatures | 10.2 | 17.1 | 26.9 | 40.3 | 51.8 | 61.7 | 66.8 | 64.2 | 54.0 | 42.0 | 28.6 | 17.4 | 40.1 |
| Precipitation | 0.77 | 0.91 | 1.91 | 2.94 | 4.33 | 4.08 | 3.62 | 4.10 | 3.50 | 2.09 | 1.32 | 0.77 | 30.34 |
| **Las Vegas, NV** | | | | | | | | | | | | | |
| Maximum Temperatures | 56.0 | 62.4 | 68.3 | 77.2 | 87.4 | 98.6 | 104.5 | 101.9 | 94.7 | 81.5 | 66.0 | 57.1 | 79.6 |
| Minimum Temperatures | 33.0 | 37.7 | 42.3 | 49.8 | 59.0 | 68.6 | 75.9 | 73.9 | 65.6 | 53.5 | 41.2 | 33.6 | 52.8 |
| Precipitation | 0.50 | 0.46 | 0.41 | 0.22 | 0.20 | 0.09 | 0.45 | 0.54 | 0.32 | 0.25 | 0.43 | 0.32 | 4.19 |
| **Reno, NV** | | | | | | | | | | | | | |
| Maximum Temperatures | 44.8 | 51.1 | 55.8 | 63.3 | 72.3 | 81.8 | 91.3 | 88.7 | 81.4 | 70.0 | 55.6 | 46.2 | 66.9 |
| Minimum Temperatures | 19.5 | 23.5 | 25.4 | 29.4 | 36.9 | 43.0 | 47.7 | 45.2 | 38.9 | 30.5 | 23.8 | 18.9 | 31.9 |
| Precipitation | 1.24 | 0.95 | 0.74 | 0.46 | 0.74 | 0.34 | 0.30 | 0.27 | 0.30 | 0.34 | 0.60 | 1.21 | 7.49 |

**Albuquerque, NM**

|  | Jan | Feb | Mar | Apr | May | Jun | Jul | Aug | Sep | Oct | Nov | Dec | Annual |
|---|---|---|---|---|---|---|---|---|---|---|---|---|---|
| Maximum Temperatures | 47.2 | 52.9 | 60.7 | 70.6 | 79.9 | 90.6 | 92.8 | 89.4 | 83.0 | 71.7 | 57.2 | 48.0 | 70.3 |
| Minimum Temperatures | 22.3 | 25.9 | 31.7 | 39.5 | 48.6 | 58.4 | 64.7 | 62.8 | 54.9 | 43.1 | 30.7 | 23.2 | 42.1 |
| Precipitation | 0.41 | 0.40 | 0.52 | 0.40 | 0.46 | 0.51 | 1.30 | 1.51 | 0.85 | 0.86 | 0.38 | 0.52 | 8.12 |

**Buffalo, NY**

|  | Jan | Feb | Mar | Apr | May | Jun | Jul | Aug | Sep | Oct | Nov | Dec | Annual |
|---|---|---|---|---|---|---|---|---|---|---|---|---|---|
| Maximum Temperatures | 30.0 | 31.4 | 40.4 | 54.4 | 65.9 | 75.6 | 80.2 | 78.2 | 71.4 | 60.2 | 47.0 | 35.0 | 55.8 |
| Minimum Temperatures | 17.0 | 17.5 | 25.6 | 36.3 | 46.3 | 56.4 | 61.2 | 59.6 | 52.7 | 42.7 | 33.6 | 22.5 | 39.3 |
| Precipitation | 3.02 | 2.40 | 2.97 | 3.06 | 2.89 | 2.72 | 2.96 | 4.16 | 3.37 | 2.93 | 3.62 | 3.42 | 37.52 |

**New York (Central Park), NY**

|  | Jan | Feb | Mar | Apr | May | Jun | Jul | Aug | Sep | Oct | Nov | Dec | Annual |
|---|---|---|---|---|---|---|---|---|---|---|---|---|---|
| Maximum Temperatures | 38.0 | 40.1 | 48.6 | 61.1 | 71.5 | 80.1 | 85.3 | 83.7 | 76.4 | 65.6 | 53.6 | 42.1 | 62.2 |
| Minimum Temperatures | 25.6 | 26.6 | 34.1 | 43.8 | 53.3 | 62.7 | 68.2 | 67.1 | 60.1 | 49.9 | 40.8 | 30.3 | 46.9 |
| Precipitation | 3.21 | 3.13 | 4.22 | 3.75 | 3.76 | 3.23 | 3.77 | 4.03 | 3.66 | 3.41 | 4.14 | 3.81 | 44.12 |

**Cleveland, OH**

|  | Jan | Feb | Mar | Apr | May | Jun | Jul | Aug | Sep | Oct | Nov | Dec | Annual |
|---|---|---|---|---|---|---|---|---|---|---|---|---|---|
| Maximum Temperatures | 32.5 | 34.8 | 44.8 | 57.9 | 68.5 | 78.0 | 81.7 | 80.3 | 74.2 | 62.7 | 49.3 | 37.5 | 58.5 |
| Minimum Temperatures | 18.5 | 19.9 | 28.4 | 38.3 | 47.9 | 57.2 | 61.4 | 60.5 | 54.0 | 43.6 | 34.3 | 24.6 | 40.7 |
| Precipitation | 2.47 | 2.20 | 2.99 | 3.32 | 3.30 | 3.49 | 3.37 | 3.38 | 2.92 | 2.45 | 2.76 | 2.75 | 35.40 |

**Oklahoma City, OK**

|  | Jan | Feb | Mar | Apr | May | Jun | Jul | Aug | Sep | Oct | Nov | Dec | Annual |
|---|---|---|---|---|---|---|---|---|---|---|---|---|---|
| Maximum Temperatures | 46.6 | 52.2 | 61.0 | 71.7 | 79.0 | 87.6 | 93.5 | 92.8 | 84.7 | 74.3 | 59.9 | 50.7 | 71.2 |
| Minimum Temperatures | 25.2 | 29.4 | 37.1 | 48.6 | 57.7 | 66.3 | 70.6 | 69.4 | 61.9 | 50.2 | 37.6 | 29.1 | 48.6 |
| Precipitation | 0.96 | 1.29 | 2.07 | 2.91 | 5.50 | 3.87 | 3.04 | 2.40 | 3.41 | 2.71 | 1.53 | 1.20 | 30.89 |

**Eugene, OR**

|  | Jan | Feb | Mar | Apr | May | Jun | Jul | Aug | Sep | Oct | Nov | Dec | Annual |
|---|---|---|---|---|---|---|---|---|---|---|---|---|---|
| Maximum Temperatures | 46.3 | 51.4 | 55.0 | 60.5 | 67.2 | 74.2 | 82.6 | 81.3 | 76.4 | 64.6 | 52.8 | 47.3 | 63.3 |
| Minimum Temperatures | 33.8 | 35.5 | 36.5 | 38.7 | 42.9 | 48.0 | 51.0 | 51.1 | 47.7 | 42.0 | 37.8 | 35.3 | 41.7 |
| Precipitation | 8.39 | 5.12 | 5.11 | 2.76 | 1.97 | 1.24 | 0.27 | 0.95 | 1.45 | 3.47 | 6.82 | 8.49 | 46.04 |

**Portland, OR**

|  | Jan | Feb | Mar | Apr | May | Jun | Jul | Aug | Sep | Oct | Nov | Dec | Annual |
|---|---|---|---|---|---|---|---|---|---|---|---|---|---|
| Maximum Temperatures | 44.3 | 50.4 | 54.5 | 60.2 | 66.9 | 72.7 | 79.5 | 78.6 | 74.2 | 63.9 | 52.4 | 46.4 | 62.0 |
| Minimum Temperatures | 33.5 | 36.0 | 37.4 | 40.6 | 46.4 | 52.2 | 55.8 | 55.8 | 51.1 | 44.6 | 38.6 | 35.4 | 44.0 |
| Precipitation | 6.16 | 3.93 | 3.61 | 2.31 | 2.08 | 1.47 | 0.46 | 1.13 | 1.61 | 3.05 | 5.17 | 6.41 | 37.39 |

**Pittsburgh, PA**

|  | Jan | Feb | Mar | Apr | May | Jun | Jul | Aug | Sep | Oct | Nov | Dec | Annual |
|---|---|---|---|---|---|---|---|---|---|---|---|---|---|
| Maximum Temperatures | 34.1 | 36.8 | 47.6 | 60.7 | 70.8 | 79.1 | 82.7 | 81.1 | 74.8 | 62.9 | 49.8 | 38.4 | 58.9 |
| Minimum Temperatures | 19.2 | 20.7 | 29.4 | 39.4 | 48.5 | 57.1 | 61.3 | 60.1 | 53.3 | 42.1 | 33.3 | 24.3 | 40.7 |
| Precipitation | 2.86 | 2.40 | 3.58 | 3.28 | 3.54 | 3.30 | 3.83 | 3.31 | 2.80 | 2.49 | 2.34 | 2.57 | 36.30 |

| | JAN | FEB | MAR | APR | MAY | JUN | JUL | AUG | SEP | OCT | NOV | DEC | ANNL |
|---|---|---|---|---|---|---|---|---|---|---|---|---|---|
| **Memphis, TN** | | | | | | | | | | | | | |
| Maximum Temperatures | 48.3 | 53.0 | 61.4 | 72.9 | 81.0 | 88.4 | 91.5 | 90.3 | 84.3 | 74.5 | 61.4 | 52.3 | 71.6 |
| Minimum Temperatures | 30.9 | 34.1 | 41.9 | 52.2 | 60.9 | 68.9 | 72.6 | 70.8 | 64.1 | 51.3 | 41.1 | 34.3 | 51.9 |
| Precipitation | 4.61 | 4.33 | 5.44 | 5.77 | 5.06 | 3.58 | 4.03 | 3.74 | 3.62 | 2.37 | 4.17 | 4.85 | 51.57 |
| **Brownsville, TX** | | | | | | | | | | | | | |
| Maximum Temperatures | 69.7 | 72.5 | 77.5 | 83.2 | 87.0 | 90.5 | 92.6 | 92.8 | 89.8 | 84.4 | 77.0 | 71.9 | 82.4 |
| Minimum Temperatures | 50.8 | 53.0 | 59.5 | 66.6 | 71.3 | 74.7 | 75.6 | 75.4 | 73.1 | 66.1 | 58.3 | 52.6 | 64.8 |
| Precipitation | 1.25 | 1.55 | 0.50 | 1.57 | 2.15 | 2.70 | 1.51 | 2.83 | 5.24 | 3.54 | 1.44 | 1.16 | 25.44 |
| **Dallas–Fort Worth, TX** | | | | | | | | | | | | | |
| Maximum Temperatures | 54.0 | 59.1 | 67.2 | 76.8 | 84.4 | 93.2 | 97.8 | 97.3 | 89.7 | 79.5 | 66.2 | 58.1 | 76.9 |
| Minimum Temperatures | 33.9 | 37.9 | 44.9 | 55.0 | 62.9 | 70.8 | 74.7 | 73.7 | 67.5 | 56.3 | 44.9 | 37.4 | 55.0 |
| Precipitation | 1.65 | 1.93 | 2.42 | 3.63 | 4.27 | 2.59 | 2.00 | 1.76 | 3.31 | 2.47 | 1.76 | 1.67 | 29.46 |
| **Houston, TX** | | | | | | | | | | | | | |
| Maximum Temperatures | 61.9 | 65.7 | 72.1 | 79.0 | 85.1 | 90.9 | 93.6 | 93.1 | 88.7 | 81.9 | 71.6 | 65.2 | 79.1 |
| Minimum Temperatures | 40.8 | 43.2 | 49.8 | 58.3 | 64.7 | 70.2 | 72.5 | 72.1 | 68.1 | 57.5 | 48.6 | 42.7 | 57.4 |
| Precipitation | 3.21 | 3.25 | 2.68 | 4.24 | 4.69 | 4.06 | 3.33 | 3.66 | 4.93 | 3.67 | 3.38 | 3.66 | 44.76 |
| **Salt Lake City, UT** | | | | | | | | | | | | | |
| Maximum Temperatures | 37.4 | 43.7 | 51.5 | 61.1 | 72.4 | 83.3 | 93.2 | 90.0 | 80.0 | 66.7 | 50.2 | 38.9 | 64.0 |
| Minimum Temperatures | 19.7 | 24.4 | 29.9 | 37.2 | 45.2 | 53.3 | 61.8 | 59.7 | 50.0 | 39.3 | 29.2 | 21.6 | 39.3 |
| Precipitation | 1.35 | 1.33 | 1.72 | 2.21 | 1.47 | 0.97 | 0.72 | 0.92 | 0.89 | 1.14 | 1.22 | 1.37 | 15.31 |
| **Seattle (co), WA** | | | | | | | | | | | | | |
| Maximum Temperatures | 45.3 | 50.1 | 52.6 | 58.3 | 64.8 | 69.0 | 74.6 | 73.6 | 69.2 | 60.4 | 51.3 | 46.9 | 59.7 |
| Minimum Temperatures | 35.9 | 38.2 | 38.8 | 42.4 | 47.7 | 53.0 | 56.0 | 56.3 | 52.9 | 47.1 | 41.1 | 38.1 | 45.6 |
| Precipitation | 5.94 | 4.20 | 3.70 | 2.46 | 1.66 | 1.53 | 0.89 | 1.38 | 2.03 | 3.40 | 5.36 | 6.29 | 38.84 |
| **Spokane, WA** | | | | | | | | | | | | | |
| Maximum Temperatures | 31.3 | 39.0 | 46.2 | 56.7 | 66.1 | 74.0 | 84.0 | 81.7 | 72.4 | 58.3 | 41.4 | 34.2 | 57.1 |
| Minimum Temperatures | 20.0 | 25.7 | 29.0 | 34.9 | 42.5 | 49.3 | 55.3 | 54.3 | 46.5 | 36.7 | 28.5 | 23.7 | 37.2 |
| Precipitation | 2.47 | 1.61 | 1.36 | 1.08 | 1.38 | 1.23 | 0.50 | 0.74 | 0.71 | 1.08 | 2.06 | 2.49 | 16.71 |

### Green Bay, WI

| | | | | | | | | | | | | | |
|---|---|---|---|---|---|---|---|---|---|---|---|---|---|
| Maximum Temperatures | 22.5 | 26.9 | 37.0 | 53.7 | 66.6 | 76.2 | 80.9 | 78.7 | 69.8 | 58.5 | 42.0 | 28.5 | 53.4 |
| Minimum Temperatures | 5.4 | 8.7 | 20.1 | 33.6 | 43.5 | 53.1 | 58.1 | 56.3 | 47.9 | 38.2 | 26.3 | 13.0 | 33.7 |
| Precipitation | 1.19 | 1.05 | 1.90 | 2.70 | 3.13 | 3.17 | 3.25 | 3.16 | 3.17 | 2.10 | 1.76 | 1.42 | 28.00 |

### Milwaukee, WI

| | | | | | | | | | | | | | |
|---|---|---|---|---|---|---|---|---|---|---|---|---|---|
| Maximum Temperatures | 26.0 | 30.1 | 39.2 | 53.5 | 64.8 | 75.0 | 79.8 | 78.4 | 71.2 | 59.9 | 44.7 | 32.0 | 54.6 |
| Minimum Temperatures | 11.3 | 15.8 | 24.9 | 35.6 | 44.7 | 54.7 | 61.1 | 60.2 | 52.5 | 41.9 | 29.9 | 18.2 | 37.6 |
| Precipitation | 1.64 | 1.33 | 2.58 | 3.37 | 2.66 | 3.59 | 3.54 | 3.09 | 2.88 | 2.25 | 1.98 | 2.03 | 30.94 |

### Casper, WY

| | | | | | | | | | | | | | |
|---|---|---|---|---|---|---|---|---|---|---|---|---|---|
| Maximum Temperatures | 32.5 | 37.4 | 43.4 | 54.9 | 66.2 | 78.1 | 87.1 | 84.8 | 74.2 | 61.0 | 43.9 | 35.6 | 58.3 |
| Minimum Temperatures | 11.9 | 16.3 | 20.2 | 29.3 | 38.9 | 47.6 | 54.7 | 52.9 | 42.5 | 33.2 | 21.9 | 15.7 | 32.1 |
| Precipitation | 0.50 | 0.56 | 0.99 | 1.51 | 2.13 | 1.24 | 1.06 | 0.63 | 0.76 | 0.88 | 0.66 | 0.51 | 11.43 |